Mathematics in a Changing World

MICHAEL HOLT & D T E MARJORAM

Mathematics
in a
Changing World

WALKER AND COMPANY
New York

First published in the United States of America in
1973 by the Walker Publishing Company, Inc.

ISBN: 0-8027-0405-0

Library of Congress Catalog Card Number: 72-95756

Printed in the United States of America.

Preface

Mathematics is making a big impact on our society—and vice versa. A generation ago, besides mathematicians, only physicists and engineers had any need of mathematics. Nowadays the picture has changed dramatically. Any sixth-former aspiring to study biology, economics, geography, sociology, and even psychology drops mathematics at his peril. And the way things are going no self-respecting language expert—whether he studies Chomsky's mathematical theory of the structure of language or not—would dare to be innumerate. Politicians, corporation lawyers, and industrial magnates must all at least be able to read a graph in their daily decision-making; doctors to talk to computers on 'hot lines' to their hospitals; farmers to interpret the growth curves of their fatstock; ethnographers to create mathematical models of primitive societies; biblical analysts to assign authorship to disputed passages in the Scriptures; and architects to design low-cost living spaces. All need some, if not very much, mathematics.

The intelligent layman, if he is to make rational decisions and vote prudently, will need at least an intuitive grasp of the economic, social, and political principles involved and these, in the last analysis, are becoming increasingly more mathematical. It is almost as though the earth shrinks to a global village, while knowledge explodes to uncontrollable proportions—as though the only way knowledge can be handled, comprehended, and communicated is by encapsulating it in formulae and symbols. It really seems that the only way to cope with our hard-won 'data' as statisticians call their figures, is for it to implode, like a popped electric light bulb, into microdot patterns in computer stores.

Now symbols have never presented any problem to mathematicians, but to the cultured layman—and perhaps to not a few scientists—mathematical symbolism forms a serious barrier to communication. In writing this book we have tried to breach this barrier. Thus although there is a scatter of signs, symbols, equations, integrals, matrices, graphs, and other diagrams throughout the text, the non-mathematical reader will find that he can skip these without losing the thread of the argument. To this end also we have cast our net wide, selected topics as much for their immediate interest as for any mathematical significance, and avoided the purely historical approach which has been done well and often before. In no sense is the book comprehensive—nor could it be when even a top flight mathematician can rarely hope to know much more than his own narrow branch of mathematics out of the many.

We live, then, in a changing world which is changing and being changed by mathematics. We have tried to convey this message by discussing, in the first part of the book, how mathematicians think, and in the latter part of the book describing what they think about. Our first respectful look is at the minds of such master mathematician-scientists as Galileo, Hardy, Poincaré, and Einstein; this leads to thinking and reasoning in general, learning processes in children and animals, and how these cognitive processes relate to logic and psychology. After a brisk overview of several mathematical models—of genetics, colour vision, drug absorption by the body, costing a published book, and plotting orbiting space-craft—we take a longer look at the details, greatly simplified, of Professor Zeeman's ingenious model of the mind, a fusion of mathematics and physiology. Next we sightsee in three great areas of mathematics in action—money, places, and people. In the long run, most arguments come down to economic ones, whether one is seeking best value or equating time with money. Today time is money.

In geography today one finds the widest and most varied use of mathematics, ranging from map projections—the most mathematical of all operations—through the statistics of settlement patterns of villages and towns, to the analysis of surfaces, shapes, and networks.

Finally, we take a fearful glance at the greatest problem of all—overpopulation. Our mathematical models today may predict population growth better than Malthusian ones, but we still do not know how to put such solutions into effect in ways that are humanly acceptable. The changing world is changing our mathematics. But in this field can our mathematics also change the world? And if so, change it in time?

To all who have helped us, materially and with encouragement, we offer our sincere thanks. In particular we are grateful to Professor E. C. Zeeman for his ready help in our popularized version of his model of the mind, and for his approval of the final exposition; any errors in it are entirely ours and not his. We are also grateful to Professor Z. P. Dienes for allowing us access to unpublished results from his researches into psychomathematics and linguistics. We warmly appreciate the creative editorial work of both Mrs Marian Kelly and Miss Marian Miller.

M.H.
D.T.E.M.

1972

Contents

CHAPTER ONE

The Mind of the Mathematician

'There are things which seem incredible to most men who have not studied mathematics.' ARCHIMEDES

'The mathematicians who are merely mathematicians reason correctly, but only when everything has been explained to them in terms of definitions and principles. Otherwise they are limited and insufferable, for they reason correctly only when they are dealing with very clear principles.' BLAISE PASCAL

'In mathematics the art of posing problems is easier than that of solving them.' GEORG CANTOR

On the notion of a mathematical set: 'A set I imagine as an abyss.' GEORG CANTOR

The nature of mathematics

The nature of mathematics almost defies definition. The apparently simple question 'What is mathematics?' has defeated many. Characteristically Bertrand Russell supplied an answer when he wrote that the whole of mathematics can be summed up in the phrase 'if p then q', implying that either not-p or q is always true, where p and q are logical propositions. Even if we cannot actually pursue his elegant and involved arguments, we can at least admire the pungency of his declaration; it probably comes as close to revealing the workings of a mathematician's mind as any statement is ever likely to.

No one—least of all a mathematician—can exactly explain the creative act in mathematics. Henri Poincaré, mathematician and brother to the French premier, in his celebrated essay on the topic

Figure 1 Bertrand Russell (1872–1970)
(*Radio Times Hulton Picture Library*)

failed entirely to disclose the innermost, secret workings of the mathematician's mind. He tells us only when and where he experienced his moments of deep insight: one on so-called Fuchsian functions arrived as he was setting foot on a bus in Normandy, and another flashed upon his inward eye as he strolled through a village while on military service. René Descartes, another French mathematician and inventor of the graph, was in the habit of lying in bed till eleven in the morning to do his thinking. Ironically, it was being made to rise at the unearthly hour of five every morning to tutor his patroness, Queen Christina of

Figure 2 Henri Poincaré (1854–1912)
(*Radio Times Hulton Picture Library*)

Figure 3 René Descartes (1596–1650)
(*Radio Times Hulton Picture Library*)

Sweden, that killed him. There is one anecdote which admirably illustrates the mathematician in him. Another of his pupils, Princess Elizabeth of Orange—he had a weakness for royalty—solved a problem, simply enough resolved by compass and pencil, by Descartes' then revolutionary 'method', as it was called. The problem was to construct a circle that touched (was tangent to) three other, quite random circles. Such a problem does not lend itself to the Cartesian method of setting up the algebra of the situation: yet Elizabeth managed to do it. Descartes' reaction was typical of a mathematician. He declared he would never undertake such a task 'in a month': too much like sheer brute slog to his mathematical mind!

Two unarguable facts about mathematicians stand out. To them, mathematics is an aesthetic and creative activity; and, like musical flair, mathematical talent is one of the plainest, most specific of human gifts—and, incidentally, is classless and raceless. Mathematicians just *know* who has a mathematical bent and who has not. Godfrey Harold Hardy, the British pure mathematician, knew that Srinivasa Ramanujan, a young, unknown, self-taught Indian, was a mathematical genius merely from reading a catalogue of his results.

The real difficulty in answering the basic question becomes evident when we realize that mathematics is not about anything at all! It is often, mistakenly, assumed that maths is nothing more than 'another language'; and this is about as true as saying that the words in which we describe an event—a description of a holiday or a horse-race commentary, for instance—comprise the actual happening itself. The words do no more than evoke the happening; and so it is with the symbolic 'language' of mathematics. The symbols recall to the mathematician whatever it is that the symbols symbolize.

All too common a fault is to confuse the symbol with the thing it represents. Commercial advertisers actually make it their job to promote this misapprehension: the advertiser's stock-in-trade is the irrationally fascinating symbol which he knows how to deploy to his commercial benefit. 'Effective rational propaganda', Aldous Huxley declares in *A Brave New World Revisited*, 'becomes possible, only when there is a clear understanding, on the part of all concerned, of the nature of symbols and their relations to things and events symbolized. Simple-minded people tend to equate the symbol with what it stands for.' Mathematicians, at least, are not naïve in *that* respect.

Another myth about the nature of mathematics also concerns symbolism: the erroneous notion that symbols of themselves are abstract. Quite the contrary, symbols cannot of themselves be abstract: they can only symbolize, recall to mind, represent an abstraction. Of course some symbols, notably Japanese characters, are more reminiscent of what they represent than are, say, the Graeco-Roman letters in which this book is printed. However crude, a picture of a cat will call to mind the idea of cat iconically, whereas the letters c-a-t do the job effectively by being symbolic. Perversely, a mathematician could, if he chose, work with symbols formed by coloured cubes or even rocks. However, the problem of first moving heavy 'symbol objects' around and then scanning the completed strings of symbols might defeat a major object of mathematical symbolization—to lighten the load on the mathematician's short-term memory. The reader need but think of doing a long multiplication 'sum' in his head. Granted the simplicity of the mathematics, he is unlikely to be able to recall the sequence of steps in the calculation without an aide-mémoire, and written (or printed) mathematical symbols as we know them are the most effective kind devised to date. Far more cumbersome were the Roman symbols, still seen on clocks and, anachronistically, on the credit titles of films. The reader has but to attempt a simple, short multiplication such as XXLIV times XIV to see the point! (Though historians claim that the Roman method was, to the Romans, an extremely efficient one of rapid calculation.) In a nutshell, the difference between Roman numerals and our Hindu-Arabic numerals lies in the use of zero and of the place-value relations concealed within them. It is the deployment of such relations that epitomizes the day-to-day work of a mathematician. His concern over zero, in particular, shows just how far in the realms of abstraction he works. For, as Alfred North Whitehead once noted, nobody deliberately 'goes out to buy zero fish'. Zero is not a very natural concept; indeed it is a highly artificial one.

Mathematical relations involve, obviously, how numbers in a series are related; but less obviously, they can be about relations (family ties) that exist between members of a human family, ties that are usually symbolized by a genealogical 'tree', though they can be treated more algebraically. The surprise in mathematics comes when we learn that we can also handle relations *between relations*, at one, more abstract, stage removed. To take an absurdly simple example: we could state the relation between each of the neighbouring numbers in the sequence 2, 4, 6, 8, and so on. Each number—after the first—is two more than its predecessor. (For instance, 8 is two more than 6.) Now we have established a serial relation between adjacent numbers which we might call the 'two more than' relation. Not satisfied with that, a mathematician might want to establish a more generalized, and at the same time more abstract, relation between all such numbers. And he would say that they are all even numbers and suggest an operational test—that we divide any such number by two and find no remainder.

Perhaps the most fruitful relations—that is, fruitful from the point of view of reading a mathematician's mind—are logical relations and more particularly *time* relations, or consequences as we call them in real life, for they underscore our ideas on causality. To say that a mathematician thinks logically is true only up to a point, as we shall see; nevertheless, closer attention to the logical processes involved in pure reasoning will be repaid.

Logic in mathematics

To anyone who knows a little of the workings of a mathematician's mind, it may come as something of a shock to realize that rarely does he create new maths simply by juggling *axioms*. Axioms are those self-evident truths which may be chosen arbitrarily but which thereafter must serve as rules of the game. The most celebrated, almost notorious, of such axioms is Euclid's fifth postulate that parallel lines never meet. (Modern interpretations suggest otherwise but then the game of mathematics has changed somewhat since Euclid played it.)

In logic, we might take the following simple scheme of statements, admittedly trivial, but which aptly demonstrates the rules:

'If it rains, the road gets wet.'
'It is raining.'

By the logical rules of inference, we can link these statements and by inference reach the conclusion:

'The road is wet.'

A common but mistaken impression is that we may deduce from these statements their converse:

'If the road is wet, then it is raining.'

A moment's thought will show why logic, for once like life, does not permit this irrational conclusion—for the road might have been hosed to keep down dust during a dry spell! None of this looks remotely like the traditional maths. Nor is it—as yet. But if we dispense with the substance and retain the form—the very essence of abstraction—we arrive at this sentence scheme:

'If *A*, then *B*' is true.
A holds.
So *B* holds.

Whatever we care to put in place of *A* or *B*, provided it is not nonsense, the scheme remains valid. Examples of nonsense statements are 'The mountain is beauty' or 'His persistence is puce'. Lewis Carroll's poem 'Jabberwocky' in *Through the Looking-Glass* teems with them. As this small but significant example shows, neither the content of the statements nor indeed the symbols to represent these abstractions count with mathematicians: it is the relationships expressed by the symbols that matter. The mathematician's craft is to construct an interesting theory from such relationships. The Greeks were the first to develop this art, epitomized for most by Euclid's *Elements*.

Productive thinking

A survey of the literature on how mathematicians think lends force to this contention. In what is perhaps the only attempt at a comprehensive survey, Jacques Hadamard, a mathematician himself, suggests that few mathematicians think in terms of words, although there have been such notable exceptions as George Birkhoff, Norbert Wiener (the creator of cybernetics) and George Polya, the American-born Hungarian educator and mathematician. Some mathematicians, he finds, belong to the auditory type, or use visual mental pictures. In an American survey many mathematicians confessed to the deliberate use of vague images. The mathematician, like Shakespeare's poet, 'gives to airy nothing/A local habitation and a name.' Verbal thinking seems to play a minor role in mathematics, as indeed it does in most creative activities other than writing or acting.

Albert Einstein himself writes:

'The physical entities which seem to serve as elements in thought are certain signs and more or less clear images which can be "voluntarily" reproduced and combined . . . this combinatory play seems to be the essential feature in productive thought . . . The above elements are, in my case, of visual and some of muscular type.'

When Hadamard put the question 'How do you think?' to Einstein, on one point the mathematician-physicist was certain. He definitely did not think in words; he thought rather in mental pictures. His own words on the matter are illuminating. The following remarks of his are selected from a book, now out of print, *Productive Thinking* by Max Wertheimer, who describes the creative process by which Einstein reached his theory of relativity. Einstein said: 'These thoughts did not come in any verbal formulation. I very rarely think in words at all. A thought comes, and I may try to express it in words afterwards.' And he laughed at the notion that, for many, their thinking is always in words. We get as close as we ever will to the method a mathematician uses in creating ideas when we accept the recommendations of Bourbaki. An intellectual jape, on an international scale, Bourbaki is the collective pseudonym of a group of some of the most creative modern mathematicians working today. In the collected works, usually known eponymously as Bourbaki, readers are exhorted to think in 'mental pictures'. What, however, does not seem to be officially recognized is the necessity to jot something down on paper to kindle the creative process. The mere act of putting pencil to paper seems to have a catalytic effect on most mathematicians' thought processes. Perhaps the most surprising aspect of mathematics in a modern world is that the basic tools of the art have hardly changed in two millenia. True, Archimedes scratched theorems in sand or fingered them in his after-bath oil on his body whereas a number theorist today might have his brainchild checked through on a computer. The scribing instrument—whether stylus and mud tablet or pencil and paper—is still the best catalyst for thought.

Traditionally, mathematics is linked in the minds of many with music; often a talent for the first goes with the second—not the other way round—but this may be no more than a reflection of a mathematician's auto-conditioned persistence, probably the most valuable attribute he can have! Certainly, a mathematician must practise his 'sums'—be they with letters and symbols in higher analysis, modern abstract algebra or straightforward higher calculus rather than with numbers—as assiduously as any musician must practise on his instrument. This fact explodes in the same breath the fondly held belief that, because mathematics is an aesthetic occupation, its beauty is open for anybody to appreciate. The truth of the matter is that, though mathematical truth may be beauty, it can only be glimpsed after much hard thinking. Mathematics is difficult for many human minds to grasp because of its hierarchical structure: one thing builds on another and depends on it. In architectural terms, the fourth floor cannot be built with logarithms till the foundations have been laid with ordinary numbers. This is not to say that certain branches of mathematics cannot be studied in isolation. Indeed it is doubtful, Polya thinks, if a mathematician today can be in any sort of creative relationship with more than a quarter of all the mathematics now known, and it is only a favoured few who are able to do so much as this.

In the authors' opinion, the nearest approach to clarifying the mathematician's kind of thinking is Wertheimer's. He sees 'productive thinking', to use his term, as quite distinct from the classical views of the nature of thoughts—traditional logic and association theory. Briefly, traditional logic was born out of Aristotle's *Organon*, was revived during the Renaissance with the idea of 'induction' from experiment and experience, and reached its apotheosis in John Stuart Mill's canon of rules for induction. Association theory, based on the work of behaviourists such as Ivan Pavlov, Edward Thorndike, and J. B. Watson, amounted to the idea of thinking in chains. (The behaviourists' critics might interpret 'chains' to mean 'shackles' rather than the intended meaning of 'strings of inferences'.) The associationist school was overthrown by the *gestalt*, or 'wholist' school, under the impact of ethologist Wolfgang Köhler's pioneer studies of apes on the island of Tenerife. This kind of work established maze-running as the cliché intelligence test for animals, and it is still accepted as such today, though as Bertrand Russell drily remarked: 'Sir Isaac Newton himself could not have got out of Hampton Court maze by any method but trial and error.' Certainly not by superior intelligence.

Today the consensus is in favour of retaining the associationist's theory to explain psychological phenomena inexplicable on any other basis. The *gestalt* school did not succeed, as Arthur Koestler has elaborately pointed out, in offering any valid alternative of its own for associative memory. (How else would we explain how a visual percept— perhaps a statue of Wellington—will summon up in the minds of many the word 'Waterloo'?) Nor has behaviourism lasted unscathed, being now regarded—and rightly in the authors' view—as 'flat-earth' psychology.

Be that as it may, within his *gestalt* interpretation of thinking in a mathematical context, Wertheimer isolates two distinct modes of reasoning—what he calls an 'A' type of response, where the behaviour is determined by the requirements of the given situation, and a 'B' response where the reaction is ordained, so to speak, by external details. As an example of a 'B' response he cites the blind recall of the rule of 'dropping a perpendicular' to establish the area of a parallelogram. Typical of a non-mathematical approach is the thoughtless repetition, developed by mindless drill, which characterizes the 'B' response—a blind, piecemeal attitude frequently inflamed by worry, personal problems and distractions. Obviously, 'A' reactions lead to the successful solution.

Wertheimer offers an illuminating, if disturbing, touchstone of the educational tenor of a school: to present a class of children with a needlessly cumbersome, 'B'-type formula such as

$$A = b\sqrt{\{(c + a)(c - a)\}}$$

for the area of a parallelogram. This formula derives conventionally enough from the more usual $A = bh$ by the simple application of

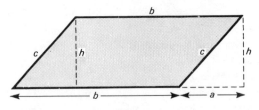

Figure 4

Pythagoras' theorem—that in a right-angled triangle the square drawn on the hypotenuse, the longest side, has the same area as the combined areas of the two squares drawn on the other two sides (see Figure 4):

$$h^2 = c^2 - a^2$$
or
$$h = \sqrt{(c^2 - a^2)}$$
$$= \sqrt{\{(c + a)(c - a)\}}$$
$$\text{Area} = bh = b\sqrt{\{(c + a)(c - a)\}}$$

In an educationally healthy school, children may recoil in disdain or even laugh; but if they accept the ridiculous formula with indifference or equanimity, the indications are that something is rotten! And, in such an educational situation, pupils are likely to be dumbfounded if invited to prove this conjecture: such an invitation touches a raw nerve of the school system, questions its very *raison d'être*. At the same time it crystallizes what mathematics is *not*.

Galileo's mind

A classic instance of 'productive thinking', typical of the way a mathematician's (or scientist's) mind works, is the genesis of Galileo's physics. The following argument comes from Wertheimer's book. Galileo Galilei, the first major experimental scientist, based most of his work—paradoxically enough—on pure thought of the 'A' type we have just discussed. The undoubtedly apocryphal story of his dropping weights from the top of the Tower of Pisa, already 'leaning' in the 1580s, obscures the strength of his contribution to mathematics and science. To see his work in perspective, imagine the prevailing intellectual climate: Aristotle's physics was unchallenged—in particular, his apparently unquestionable law that a 'moving body sooner or later comes to rest if the force pushing it no longer acts'. Galileo's thinking provided, like St Augustine's some 1200 years before, a bridge from the old world to the new, this time an intellectual construct, where St Augustine's had been a spiritual one. Indeed, it was by way of Galileo's

Figure 5 The Tower of Pisa from an early engraving
(*Radio Times Hulton Picture Library*)

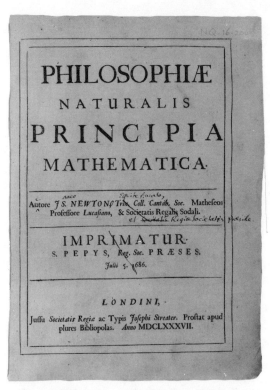

Figure 6 The title page of Newton's *Principia Mathematica*, annotated by the author (*Trinity College, Cambridge*)

bridge of abstractions that Newton produced his work of genius, the *Principia*, which embodied his universal law of gravitation and which was to remain unchallenged until Einstein propounded his general theory of relativity. Let us look closer at Galileo's powerful way of thinking.

The facts were, as anyone can verify: when a stone is thrown up in the air it *decelerates* (slows down) in its upward flight until it hovers at its zenith before plummeting to earth once more. Galileo knew that the acceleration due to the Earth's gravitational attraction was constant and always directed towards the Earth's centre, but he also discovered that a ball rolls down a sloping ramp with a smaller acceleration. 'Necessity is the mother of invention', and he had been forced to 'dilute' gravity with the help of a sloping ramp. It is a matter of everyday observation that a cyclist picks up speed faster (accelerates more) when free-wheeling down a steep hill than down a gentle incline. Contrariwise, as a mathematical thinker *would* think, a marble will lose speed more rapidly when impelled up a steep ramp instead of up a gentler one. Galileo abstracted the facts of the situation something on these lines.

Having productively disentangled his thoughts from the irrelevancies of the real world, Galileo was free to ponder the situation in the abstract and from there perhaps to generalize to a wider situation. From being thrown straight up to being rolled up a gentle incline, the marble experiences a slackening of retardation. To put it another way, the gentler is its incline, the less the ramp dilutes the pull of gravity. The same argument, in reverse, applies to dropping a marble or letting it fall down the ramp. The dynamic symmetry of the mental picture of the situation (see Figure 7) leads ineluctably to the question: 'What happens when the ramp is level?' Despite all physical evidence to

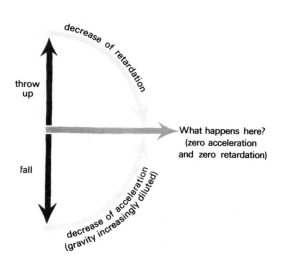

Figure 7

the contrary and notwithstanding Aristotle's impeccable and time-honoured reputation, Galileo theorized that a body moving on the level will continue to move until stopped by some outer force. The gigantic measure of his intellectual leap can be gauged when we recall that nothing evades the drag of friction—not even a chunk of ice slipping across a frozen lake. Galileo's thought experiment quite transcended everyday reality.

Two episodes in Galileo's work illuminate the simplicity yet depth of his mathematical thinking. The first concerns his experiments with a ball descending a grooved and polished inclined plane. In his laboratory at Pisa, he and an assistant set up the plane and a water clock. As Galileo describes it:

'. . . we let the ball roll down the channel, noting . . . the time required for the descent . . . we now let the ball roll down only one quarter of the length of the channel, and having measured its time of descent, we found it to be precisely one half of the former . . . In such experiments, repeated a full hundred times, we always found that the distances traversed were to each other as the squares of the times.'

Galileo does not explain this squared relationship; he merely states how the ball rolled down. In a later report he records his discovery that 'the distances traversed, during equal intervals of time, by a body falling from rest, stand to one another in the same ratio as the odd numbers beginning with the unity . . .' It is hard to doubt that he was linking the squared numbers $1 (= 1 \times 1)$, $4 (= 2 \times 2)$, $9 (= 3 \times 3)$, $16 (= 4 \times 4)$, $25 (= 5 \times 5)$, and so on with their differences, the odd numbers: 3, 5, 7, 9, and so forth. The Greeks had pictured this connection between the sequences with their so-called figurate numbers (see Figure 8).

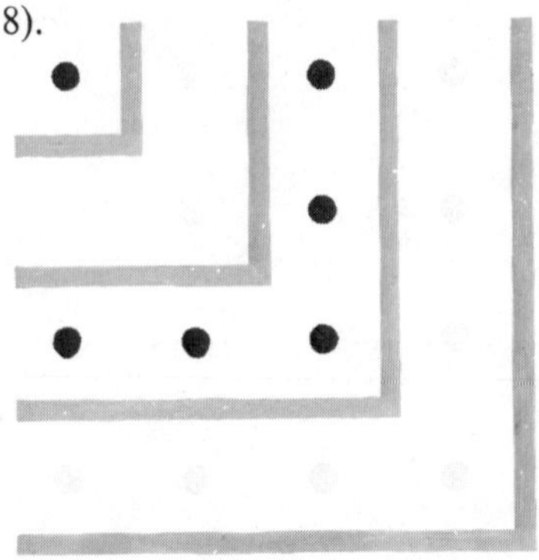

Figure 8 Figurate numbers

The second episode reveals Galileo's understanding of that most mathematical of concepts, infinity. In his *Dialogue on Two New Sciences* he has three imaginary characters: Salviati, who was not unlike himself, Sagredo, an intelligent layman who 'sits on the fence', and Simplicio, a simple-minded, purblind follower of Aristotle, who was, according to some scholars, modelled on Lodovico delle Colombe, a Florentine scientist who sought to discredit Galileo. The Jesuit, Christolph Scheiner, who claimed to have preceded Galileo in the discovery of sunspots and with whom Galileo was already on bad terms, seized the opportunity to make mischief. He asserted that the simpleton was a pen-portrait of His Holiness, Urban VIII. Galileo actually has Simplicio repeat the Pope's escape clause for the believer with a scientific 'conscience'—that a hypothesis may explain the facts satisfactorily, but God may have produced the same phenomenon by different means—thus publicly cocking a snook at the Pope. In effect, Galileo asked for a showdown, and he got a trial!

Galileo has Simplicio pose a (then) unsolvable riddle; a long line segment has an infinity of points but then so has a short line segment. The dialogue runs:

Simplicio: '. . . the infinity of points in the long line segment is greater than the infinity of points in the short line segment. This assigning to an infinite quantity a value greater than infinity is quite beyond my comprehension.'

Indeed it was to remain beyond man's comprehension till the end of the nineteenth century when the Russian–Danish mathematical genius, Georg Cantor, resolved the problem with the theory of the infinite. Like Galileo, Cantor suffered for his ideas: he was barred from promotion in all German universities as a result of his publication, and mathematical journals refused to publish his papers. But, unlike Galileo, he was reconciled with the Church, and he received praise from quite an unexpected direction—the Jesuits lauded his theory. They saw in it irrefutable proof of the existence of God!

In Salviati's reply to Simplicio, Galileo comes to the rescue.

Salviati: 'This is one of the difficulties which arise when we attempt, with our finite minds, to discuss the infinite, assigning to it those properties which we give to the finite and limited; but this is wrong, for we cannot speak of infinite quantities as being the one greater or less than or equal to another. To prove this I have in mind an argument which, for sake of clearness, I shall put in the form of questions to Simplicio who raised this difficulty. I take it for granted that you know which of the numbers are squares and which are not.'

Here Salviati was referring to the counting or natural numbers, excluding zero, of course. Galileo has him show that there are as many square numbers 1, 4, 9, 16, 25 and so on, as whole numbers, 1, 2, 3, 4, 5 and so on. Bafflingly he infers that there are, so far as he can see, as many numbers altogether as there are squares—though the squares are among the whole numbers anyway. It fell to Cantor to cut this Gordian knot. He did so by invoking the concept of a *set*—broadly speaking, a collection or array of distinct things. (The idea of a set and the associ- ated language for describing sets are now taught as the linchpin of the 'new maths' in most secondary schools in America, Britain, and many European countries.)

To return to Galileo's thought experiment—it is a prime example of the use of a mathematical 'model'. Most of modern theoretical physics and much modern mathematics are advanced today by the device of 'model-making'. It should, perhaps, be pointed out that such a 'model' is not a concrete structure like a skeleton or a mobile—though, for instance, an Arthur Caldwell mobile might well serve one day as a physical representation for such a mental model. A mathematical model is a symbol for a piece of imagery, a mental picture impaled on paper like a collector's butterfly. Galileo's particular model highlighted the crucial issue in the problem of moving bodies. What happens to the object moving on the level? The nub of productive thinking is to get at the crucial issue by making a predictive model.

Einstein's mind

A more recent example confirms the same aspect of productive thinking. Einstein conceived the idea of his special theory of relativity as a sixteen- year-old student. He worried at the problem for seven years, yet, once he had it solved, it took him a bare five weeks to write it up in his epoch-making paper on electromagnetism. His quest started with a question (as most research does): 'What would happen if one were to chase a ray of light?' More fancifully, he might have asked: 'If one could chase a moonbeam fast enough, would it no longer appear to move at all?' He found it puzzling, too, to conceive of a ray of light moving faster in one direction than another. Now this is not the place to retell the story of the physical discoveries that led up to his theory. It is enough to recall that early twentieth-century physics unarguably pointed to the fact that light has a constant velocity, and this meant that light was a bit like Stephen Leacock's man who 'rode madly off in all directions'—and each way at the same speed.

Why was light's constant velocity so disturbing to physicists? As one suggested, it was as if we were to say that a man walking up an escalator did not get to the top any sooner than a man standing still on the escalator. To extricate science from this embarrassing situation, an Irish physicist, George Francis Fitzgerald, advanced an audacious and, perhaps, characteristically Irish solution: a moving body shrinks along the line it moves in! Independently, a Dutch mathematical physicist, Hendrik Antoon Lorentz, worked out a more sober mathematical *invariance* relation, which gave the Fitzgerald contraction again, but alas got no nearer to the insight necessary to explain such extraordinary goings-on. In the event it fell to Einstein's genius to do just that.

Einstein's train of thought, Wertheimer reports from conversations with him, began with the presupposition that light did not move at a constant velocity—that is, did not move at the same speed in all directions. But all his efforts to reconcile this assumption with the accepted body of physics theory failed utterly. For instance, the inconstancy of the velocity of light would not fit with the bed-rock formulae of electricity and magnetism, 'Maxwell's equations' as they are known; any theory about light had to agree with the Maxwell equations because light is an electromagnetic phenomenon—that is, involving both electricity and magnetism. Finally abandoning this approach, Einstein posited that the velocity of light *was* constant. What then? He had recourse to what was to him a characteristic mode of thinking: he questioned the most seemingly obvious facts of science. In this he was inspired by certain philosophers of science. As James R. Newman says: 'He acknowledged his debt to . . . Henri Poincaré, and of course to the great Scottish sceptic of the eighteenth century, David Hume. Hume took the cheerful view that we know nothing and that there is no such thing as a rational belief.'

Put in mathematical terms, Einstein reviewed the basic axioms of the system—a tried-and-tested method for producing fruitful lines of enquiry in mathematics research. In thinking terms, he isolated the key area. He went for the time-honoured notion of simultaneity: he asked how we know two events at different places are simultaneous. He also insisted there be an acid test, an operational definition, for such a concept of simultaneity. 'The concept', he said in discussion, 'really exists for the physicist only when in a concrete case there is some possibility of deciding whether the concept is or is not applicable.' Without such a hard and fast ruling, he felt 'I am deluding myself as physicist (to be sure, as non-physicist, too!) if I believe that the assertion of simultaneity has any real meaning.'

This is not the book in which to recapitulate any of the countless 'popular' expositions on relativity. Suffice it to say that for the most part the 'popularizers' side-step one highly salient point: the meaning of 'relativity'. One of the clichés about the theory is that everything is relative. As Russell has pointed out there must be something for everything to be relative to—a fixed, non-relative thing, an absolute. As Humpty Dumpty might have put it: 'Everything relative! Why, you might just as well say everything is more (or less).' Marginally, be it noted, the word 'relativity' does not appear in the title of Einstein's epochal paper of 1908 in which he developed the theory: it was entitled *On the Electrodynamics of Moving Bodies*.

Einstein was really at pains to find an invariance—something that, far from being relative, is absolute and does not change from one system to another; he found it in the velocity of light. Richard Feynman, an American Fellow of the Royal Society and a leading theoretical physicist, has suggested that physics is but the study of invariance. What this means is that physicists want to know that their laws, for example, Newton's law of gravitation, work equally well in, say, the constellation Andromeda as here on Earth—in a word, that such laws are invariant. (We shall see the significance of this aspect in the section *A whiff of grapeshot* in Chapter 4.)

Einstein solved the problem of relativity by conceiving the remarkable idea that the velocity of light was the greatest attainable anywhere in the universe. Here was the turning point in his thinking process—a gigantic leap which led to a mathematical formulation that happily embraced Lorentz's work and placed Fitzgerald's apparently arbitrary hypothesis of contraction within a deeper, more profound context. Before his discovery that the crucial point lay in querying the hitherto unquestioned concept of simultaneity, Einstein knew when he had got to the crux of the matter. Even so, he never used the basic axioms first. 'No really productive man thinks in such a fashion,' he told Wertheimer. 'The way the two triple sets of axioms are contrasted in the Einstein–Infeld book [a classic account of Einstein's discovery written by him and a colleague] is not at all the way things happened in the process of actual thinking. This was merely a later formulation of the subject matter.' A case of hindsight gilding the scientific lily. 'The axioms express essentials in a condensed form. Once one has found such things one enjoys formulating them in that way; but in the process they did not grow out of any manipulation of axioms.'

The constructive and intuitive aspect rather than the axiomatic is stressed by two mathematicians in America, the late Richard Courant and Herbert Robbins when they write: 'If the crystallized deductive form is the goal, intuition and construction are at least the driving forces.' They suggest that if mathematics were nothing but an arid, whimsical system of axioms and conclusions, the subject 'could not attract any intelligent person. It would be a game with definitions, rules, and syllogisms, without motive or goal.' As for the axiomatic aspect, the English relativitist, C. W. Kilmister, has this to say: 'It is quite possible to imagine a system in which every proposition has a proof or disproof, but in which there is no uniform procedure for discovering proofs, so that skill and ingenuity are essential. Until 1930' —with the arrival of Gödel's theorem—'many people supposed mathematics to be such a system . . .'

The fact is, mathematics has attracted the most creative minds since the subject began in earnest in the days of ancient Egypt and Sumer. Because the subject is an epitome of all thinking, it is perhaps not surprising that the tools of mathematics—abstraction, symbolization, and generalization (of which more in Chapter 2)—have not changed much since Euclid was demonstrating that there is 'No royal road to geometry' (the top branch of mathematics in his day). It is sometimes mistakenly believed that the computer has changed the face of mathematics. Nothing could be farther from the truth: admittedly the computer has reduced much of the slog and drudgery involved in mathematics, but the curious, abstract processes involving the mind are still much as they were when Henri Poincaré wrote the following much quoted passage:

'It never happens that the unconscious work gives us the result of a somewhat long calculation *all made*, where we have only to apply fixed rules. We might think the wholly automatic subliminal self particularly apt for this sort of work, which is in a way exclusively mechanical. It seems that thinking in the evening upon factors of a multiplication we might hope to find the product ready made upon our awakening, or again that an algebraic calculation, for example a verification, would be made unconsciously. Nothing of the sort, as observation proves. All one may hope from these inspirations, fruits of unconscious work, is a point of departure for such calculations. As for the calculations themselves, they must be made in the second period of conscious work, that which follows inspiration, that in which one verifies the results of this inspiration and deduces their consequences. The rules of these calculations are strict and complicated. They require discipline, attention, will, and therefore consciousness.'

CHAPTER TWO

Thinking and Reasoning

'There is a very good saying that if triangles invented a god, they would make him three-sided.'

CHARLES, BARON DE MONTESQIEU

'Four legs good, two legs bad.'   GEORGE ORWELL

'Mathematics is cleverer than the mathematicians.'

A saying

'The understanding of atomic physics is child's play compared with the understanding of child's play.'

DAVID KRESCH

Thinking in animals

Anthropologists have discovered a mathematically remarkable fact: some Stone Age tribes, such as the Papuans of New Guinea, once head-hunters, have no concept of an alternative choice. Offer a head-hunter a choice of gifts or courses of action and he is utterly at a loss; he cannot grasp the logic of alternatives, the logical connective *'either . . . or . . .'* Perhaps this should not surprise us. At an everyday level of decision-making, choosing one of two alternatives presents most of us with a 'headache'. And few appreciate that even refraining from making a decision is in itself a decision.

Everyday language confuses the issue, too. We frequently refer implicitly to what logicians call the *exclusive* 'or'. Either we go to the Moon or we stay on Earth; we cannot go to both places at once, for logical statements of this kind are implicitly contemporaneous. In mathematics, the *inclusive* 'or' is a more useful concept. We can have

A or *B* or both. This particular way of linking two things—of (impossibly) having your cake and/or eating it—is a basic logical relation. Presently we shall look more intently at the nature of relationships.

On a more sophisticated level of appreciation, there seems to be evidence for an evolution of intelligence from *Homo faber*, the tool-user, to *Homo sapiens*, the wise one, whose intelligence philosopher Herni Bergson saw as the rational intelligence of classical psychology and logic. In an often quoted passage from his book, *Creative Evolution*, he said in 1902, almost as if he had 1969's first Man-on-the-Moon landing in mind:

'Intelligence in its initial form may be considered to be the ability to produce artificial objects, particularly tools, and to continue making them by a variety of methods. If we could ignore questions of pride, and if we could let ourselves be guided by history and prehistory alone, we should perhaps call ourselves, not *Homo sapiens*, but *Homo faber*.'

Animals display a surprising number of intelligent strategies. Zoologists J. Z. Young and B. B. Boycott demonstrated the ability of the octopus to learn and remember—two prerequisites of success at mathematics. An octopus can learn to distinguish different-sized squares, white and black circles and other geometrical shapes by sight. Wolfgang Köhler, the famous German ethologist, noted for his work with apes, also showed that chickens have an aptitude for 'seeing' relationships. He placed two sheets of grey paper, one bright and one dull, in their feeding area; at the outset he put food down on the dull grey sheet. By replacing the dull grey card with an extra-bright grey card, he discovered the fact that chickens respond to *relative* brightness, not absolute brightness. Arguably, this is evidence for an innate feel for a mathematical relationship—programmed, as it were, into animals' brains. More convincingly, Otto Koehler (not to be confused with Wolfgang Köhler) found that magpies can discriminate *at a glance* up to eight objects, whereas man, at best, can see at a glance no more than five things.

Ivan Pavlov's dogs were tested for their ability to differentiate geometric shapes. In one experiment, recorded in Pavlov's famous book *Conditioned Reflexes*, a dog was offered a dark 'T' as a positive stimulus and it could distinguish this from the other dark figures and a white-on-dark 'T' (see Figure 9). Another Pavlov dog could tell the difference between the positive stimulus of a white cross and other white symbols

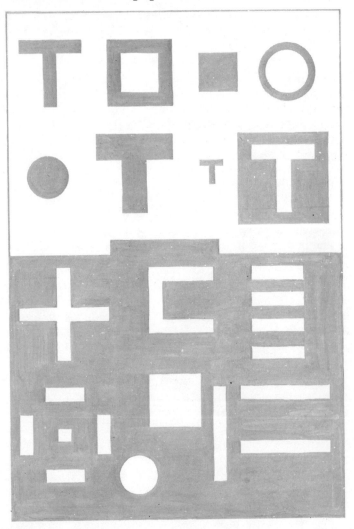

Figure 9 Examples of the geometrical shapes used by Pavlov in his differentiation experiments with dogs

also shown in Figure 9. Such discrimination is essential to mathematical thinking—indeed to any kind of thinking—for mathematics begins with classifying, differentiating, and distinguishing relationships between the classes so formed. W. H. Thorpe and other psychologists have established beyond doubt that birds have 'a pre-linguistic number

sense'; among mammals, squirrels have been shown to be able to abstract the concept of numerical equality from totally disparate objects of different size and colour. Man's advance on this pre-linguistic counting ability is to graft onto it the symbolism to record it.

Recently, in the US, psychologist Benjamin Weinstein conducted experiments on rhesus monkeys to find out if monkeys can think. The experiments, by the way, have a strong resemblance to much Montessori-inspired work done nowadays with infant and junior school-children all over the world. Weinstein first conditioned a female rhesus monkey to react to colour by teaching her to push forward black objects when shown a black triangle, and brown objects when shown a brown circle (see Figure 10). Then he went on to the more difficult step: he presented her with an uncoloured triangle and circle. The monkey was still able to select the appropriate colours, thus proving she had made the sophisticated mental leap of linking up a shape with a colour. Mathematically speaking, the monkey was *abstracting* and reading symbols.

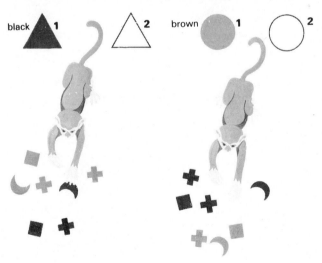

Figure 10 Weinstein's experiment. A rhesus monkey links, in her mind, shape with colour—she abstracts

Indian psychologist Sheo Dan Singh has a battery of 'oddity' (odd shape out) tests to compare the intellectual abilities of urban-raised rhesus monkeys in Jaipur, India, with their forest-dwelling cousins (see Figure 11). He observed that

'the urban way of life causes monkeys to change their feeding and sleeping habits, alters their behaviour toward one another, increases their aggressiveness, makes them highly responsive and manipulative in their approach to novel and complex features of the environment and in general enhances their psychological complexity, but *it does not advance their intelligence*, [authors' italics] although their behaviour may appear to exhibit a high degree of shrewdness. I leave it to the reader to speculate on what implications these findings may have for understanding the impact of urban life on the behaviour of man.'

Figure 11 Sheo Dan Singh's 'oddity' tests for comparing the intellectual abilities of rhesus monkeys

In a busy, congested classroom, a city child's 'slickness' may pass for real mathematical insight.

Rats have been trained to abstract in recognizing 'two-ness'. Three doors are marked with one, two, and three symbols, such as coloured triangles or circles or, even, cats! Only the door with two symbols on it leads to food, and the rat learns to jump through only this door. It finds what is common to all the elements (symbolic presentations) in the class of two-symbol presentations. Suppose for a moment that a super-intelligent rodent, such as the rat hero of the book *Flowers for Algernon* were involved in such an experiment: he might well extend this class of 'two-ness' from signs on a door to a wider class of pairs, which could employ auditory, tactile, and finally perhaps non-iconic representations with no pictorial content, such as the Hindu-Arabic numeral '2'—a straightforward case of generalization. The reverse process of particularization is a relatively easy and commonplace intellectual activity for children and adults alike.

Abstract thinking

Koestler likened the process of abstraction to forming a memory of a television play. We remember the gist of the play in the end as the story of a gangster on the run, but we have forgotten the actors' lines and the sequence of scenes. 'If this were the whole story,' Koestler continues, 'memory would be a collection of dusty abstracts, the dehydrated sediments in the wine glass, with all the flavour gone.' This may be a clue to the reason why so many people actively hate mathematics: the intellectual regimen imposed by the subject may seem to them a form of mental dehydration; indeed, the intellectual self-consciousness, part of the mathematician's make-up, is repugnant to easier-going minds. Making a distinction rather like the Piaget–Inhelder dichotomy of 'scheme' and 'schema', Koestler contrasts abstraction with the *picture-strip* type of memory, which recalls a 'detail like a wart on granny's chin, or the taste of Proust's *madeleine*, or a single gesture of an actress in an otherwise long-forgotten play'. Most seem to agree that abstraction is 'a rule of action', as F. A. Hayek calls it, and he goes so far as to suggest the novel idea of the 'primacy of the abstract' in the sense that a child abstracts first and gains experience later. (Of course, the child has the experience of the species to draw on.) His first abstraction is that things that cause pain are bad.

In contrast to the agreement about the nature of abstraction, there is no such apparent concord when it comes to defining generalization. Zoltan P. Dienes sees it as an extrapolation—almost a mental leap in the dark—from everyday experiences, whereas Karl Popper emphatically does not. Popper's argument against 'inductivism' (moving from the particular to the general) hinges on his belief that we cannot logically derive generalizations from particular experiences. The capacity to generalize comes first; hypotheses are then tested and confirmed or refuted according to their effectiveness as guides of action. Most children play at 'killing' and shout 'You're dead!' without having seen a dead person, and certainly without having experienced death! It is perhaps in the act of generalizing that the significance of purely playful activities of animal and human intelligence resides.

Inability to generalize has powerful psychological undertones. Some people experience genuine *angst* in trying to generalize—possibly an explanation of racialism, as Milton Rokeach's researches on dogmatism in the US showed. He found that people were far less ready to accept the religious values of a sect too near the beliefs of their own denomination—a case of the pain of generalizing, as opposed to the ease of thinking something totally disparate, as in the case of a Christian considering with dispassionate equanimity the very different views of, say, a Muslim. Rokeach's findings suggest far-reaching conclusions. The very imposition of external punishment, in whatever form, is conducive to the formation of a rigid, single-minded personality, to tough-mindedness, redolent of the Nazi leather-jacket mentality, as Adorno in the US showed. And, as if this were not distressing enough, 'extra maths and poetry' have, for obvious but none the less deplorable reasons (because they are the handiest, most painless correctives!), been traditionally meted out as sticks with which to punish unwilling pupils.

Today we cannot avoid being struck by the darker side of the personality coin: in totalitarian countries dictators have summarily and absurdly suppressed the teaching and banned the learning of the new mathematics —doubtless for the very good reason that people who can and do think are a greater headache to dictators than those who won't or can't. The death of Archimedes, one of the greatest mathematicians who ever lived, is poignant testimony of the leather-minded man-of-action's dislike of the thinker. During the sack of Archimedes' town Syracuse in Sicily, the sage, then seventy-five years of age, was poring over a problem drawn in the sand. Against, it must be admitted, the conquering general's express command, a soldier slew Archimedes because, one story has it, Archimedes ignored the soldier's presence. As Whitehead once wrote drily: 'No Roman ever lost his life for contemplating a problem in the sand.' The nineteenth-century historian, John William Draper, writing of the genius of Rome—that 'manifested itself rather in physical than in intellectual operations'—might have been writing about cultural changes of today: 'The Roman soldier is about to take the place of the Greek thinker, and asserts his claim to the effects of the intestate—to keep what suits him, and to destroy what he pleases.'

Everyone carries in him the dragon's teeth seeds of the Nazi mentality. 'Since the slaying of Abel', as Paul MacLean said at the 1968 Alpbach Symposium, '[Man] has had to live with the realization that his fellow man holds over him the power of life or death . . .' It puzzled him 'how a civilized people can be duped into the selection of a deranged leader . . . Particularly deceptive it seems, are the bold, aggressive qualities of the psychopath that make it possible for him to put on a big show and talk louder and longer than anyone else.'

It is to be hoped that since the outset of the humanitarian movement, the layman should be able to domesticate his emotions—perhaps, we might suggest, through a greater emphasis on 'thinking' as part of the school curriculum.

Abstraction and generalization

These terms are repeatedly used by mathematicians and physicists, and in a different sense by psychologists. Even to the mathematician their meaning is imprecise. For first we must establish, as Whitehead asserts, whether we are in the realms of possibility or actuality. The mathematician is usually working in the first realm when he builds an elaborate mental picture. Aristotle was tackling the second realm when he used logic to classify and analyse concrete facts into more abstract elements. So is a botanist when he arranges flora and fauna into a taxonomic scheme of families and species.

It may help here to picture a cone of abstraction standing with its broad base firmly on the ground of concrete actuality: here things are very complex. Then we move up the cone to the headier heights of abstraction, the complexities dwindling with the cone's girth, until we reach the summit, the quintessence of abstract simplicity and quite remote from reality. As an example, we might abstract from all the many kinds of cats in the world, each belonging to a different species—

lion, tiger, puma, and so forth—the single genus 'cat', the essence of 'catness' perhaps but not like any single cat in the world.

Some confusion of thought arises from the fact that abstraction from actuality runs in the opposite direction to abstraction from possibility, the mathematician's way. The latter may be seen as an inverted cone, its point embedded in the realm of the possible: the more abstract we become, the farther up towards the wide base of complex abstractions we move. As a result, writes Whitehead, 'the simple eternal objects represent the extreme of abstraction from an actual occasion; whereas simple eternal objects represent the minimum abstraction from the realm of abstraction.' It is this second construction we usually mean when we talk of 'abstraction'.

Before examining abstraction in mathematics, it is worth noting what happens in science. Science achieves unification by shifting to a higher order of abstraction. But at a price. 'Although the trend toward unification through abstraction is probably inevitable as science grows,' writes Alvin Weinberg, 'it is well to remember that we pay in loss of resolution for the broadened viewpoint we gain as we become abstract.'

J. Z. Young has gone some way to show the process of abstraction in real life as well as in the laboratory. Young has depicted a single memory unit—what he calls a 'mnemon' (see Figure 12). It incorporates two principles: selection of information ('abstraction') and prediction ('generalization'). A classifying cell records the occurrence of a particular event. It has two outputs, producing possible alternative motor actions. The system is biased to one of these (say, 'attack' in Figure 12).

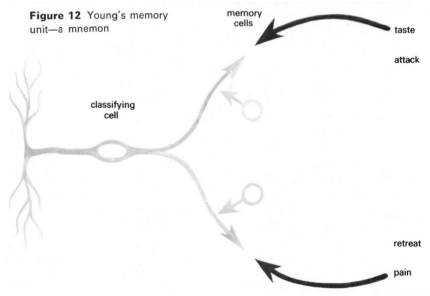

Figure 12 Young's memory unit—a mnemon

memory cells

taste

attack

classifying cell

retreat

pain

Following this action, signals initiating this action arrive and either reinforce what was done or produce the opposite action. Collaterals of the higher motor cells then activate the small cells, which produce an inhibitory transmitter and close the unused pathway. These may be called 'memory cells' because their synapses can be changed.

The principle of the situation comes to this: the cell receives a signal; it reacts one way or the other—the direction being determined by its past experience—after testing. In the light of this testing, it reacts further.

Suppose a dog reacts to a stimulus as if it were an accompaniment of 'attack'. What the stimulus really means becomes clear only after the dog has made a test. The attack may have been only a token, a form of animal shadow boxing. But how can the dog tell till it has tested by making some slight response? Under certain conditions 'attack' may be transmuted from pain to pleasure (mock attack, puppy-play, for instance). The 'pleasure' and 'pain' states are generated within the animal; they are 'self-states', or what a physicist would call 'eigen-states'. These self-states are analogous to what we call 'states of mind' in a man. And they clearly colour the dog's mode of reaction to the incoming signal. For instance, a well-disposed dog might interpret an attack signal as an invitation to play, and react accordingly, with happy or dire results, as the case may be. Or the dog might read such a signal as a serious threat and respond by raising hackles, baring teeth and snarling. The dog's interpretation of the signal is palpably a form of abstraction. After a number of mock attacks he learns to abstract the essential 'play' aspect; after several hostile threats he abstracts the 'attack in earnest' attribute of the signal. In short, the better he has abstracted, the less he is prone to be fooled.

This immunity to hoodwinking seems to be an acid test of abstraction having taken place in human beings, too. Take the anecdote of the small boy learning addition:

Teacher:
Tommy, I've got 2 pears and 3 pears. How many "fruit" have I got altogether?'
Tommy:
'Please, Miss, we've only done sums about apples.'

For Tommy, the abstractive process of adding had clearly not happened.

Again, Miranda, Holt's daughter, when 7 years old, was puzzled to learn that 100 and 3 makes a-hundred-and-three. For, as she said:

'2 and 3 is 5. It isn't two-and-three.' She had not abstracted the concept of place value—though she was clear enough in her mind to remark on the seeming discrepancy. Now at what Jean Piaget calls the pre-operatory level (before 7 to 8 years old), the child needs to make sure by actually handling objects that $3 + 2 = 2 + 3$ or that $A = C$ if $A = B$ and $B = C$ (when he cannot see A and C together)!

To develop their ability to abstract the idea of number children seem to need to work in bases other than 10, the base in which most of us count. Traditionally, arithmetic was taught using only base 10 with a little practical work grouping things in tens. Grouping in hundreds, let alone thousands, is out of the question: it would take a whole lesson to count 100 marbles with little assurance of accuracy. How then do we provide practice in handling place value with real objects? By using smaller bases—base 2, 3, 4 or 5.

In mathematical terms any number N may be written as:

$$N = a_z b^z + \ldots + a_3 b^3 + a_2 b^2 + a_1 b + a_0$$

Here b is the base one is counting in; z is the greatest power of the base used, and a_z is the actual digit written in that place. We know each of the digits a_0, a_1, a_2, a_3 up to a_z is less than b, the base. For instance, with ordinary counting numbers, the digits 0, 1, 2, . . . up to 9 are all less than the base 10.

If the formula strikes the reader as puzzling, it may be because the algebraic symbolization is strange or that the generalized notion of a number written in any base is unfamiliar.

Suppose $\qquad\qquad N = 137$
Then $\qquad\qquad N = 1 \times (10)^2 + 3 \times (10) + 7$

In this particular case, the base $b = 10$, and the digits $a_0 = 7$, $a_1 = 3$, and $a_z = a_2 = 1$. (Here z, the greatest power used, is 2.) We can generalize this expression to any number of places z as we did above. Or we could write it as follows:

	thousands	hundreds	tens	units	
. . .	0	1	3	7	in base 10
. . .	a_3	a_2	a_1	a_0	in general terms in base b

In conventional arithmetic teaching, z is at most 3, giving the thousands place. Every conceivable change of a_0, a_1, a_2, a_3 is given in the form of practice or drill exercises. Though why this is deemed so worthwhile in the classroom is not clear. Not noted for mathematical prowess, bookmakers' clerks were nevertheless legendary for being fast *and* accurate at computing odds, admittedly on a purely associative basis. Of course, their method of learning the Four Rules (of add, subtract, multiply and divide) was effective, if unmathematical. They had for the most part simply established associative links with little or no relational connections. It may be that children should be rewarded for their dull labours just as adults expect to be—with money or in kind! The ineluctable truth is that most arithmetic 'sums' amount to 'symbol shoving' and rely for the most part on 'visual' cues only— in the form of signs on paper. A striking fact (discussed later in this chapter in the section *Perception and learning*) emerges about teaching maths. Even the most superficial glance at the various modes of teaching maths—employed in classrooms all over the world—reveals that most cues for mathematics learning are almost entirely 'visual'. For instance, the noted physicists, Neils Bohr and Max Born, mention 'sensory experiences' in their discussions of theoretical physics but do not make clear that it is the use of the optic sense which they mean.

Fortunately light is, for Bohr, 'to all accounts the simplest of all physical phenomena'. The power of the verbal pun is shown by the *double-entendre* of the word 'see': a child who suddenly succeeds in doing a mathematical problem will very likely exclaim 'I see it!' though there is nothing to be seen visually. The creators of *gestalt* psychology— so powerful a tool in modern mathematical education—could scarcely have enjoyed success if they had adopted the sense of smell or hearing instead of sight.

We abstract then when we learn to handle 2 or 3 or any number of anything. And when we learn that $3 + 2 = 2 + 3$ and many more such examples, we can generalize to the formula in algebra $x + y = y + x$ where x and y each represent any number. A more powerful form of generalization is at work where we extend a class to a wider one so that the wider class contains within it a part just like the original class— as in extending the idea of the counting numbers $(1, 2, 3, \ldots)$ to the wider class of positive and negative integers $(\ldots -2, -1, 0, +1, +2, \ldots)$. The counting numbers 1, 2, 3, 4 and so on are less general than the numbers with signs $+1, +2, +3, +4$, and so on *and* $-1, -2, -3, -4$, and so on. The counting numbers behave just like the signed positive numbers but, of course, are not the same. Two dogs might be a 'couple' but would never be $+2$ dogs! The sign gives a sense to the numbers as, for instance, with $+10°C$ or $-5°C$. Again fractions are more general than whole numbers. Also adding times on a 12-hour clock can be generalized in the same way to adding times on a 24-hour clock. Some of the 12-hour clock times will be embedded in the 24-hour clock times.

'Time' provides a classic example of a generalization. 'What is time then?' St Augustine asks himself in his *Confessions*. 'If nobody asks me, I know: but if I were desirous to explain it to one that should ask me, plainly I know not.' Yet if we were to ask Augustine—or anyone else, come to that—'What's the time?', he would have little difficulty in responding. The second query is, of course, a particularization of the first; the first query a generalization of the second; and both questions deal in abstractions.

It is possible to generalize simply by playing around with symbols. The trouble is, however, that students are often given to doing this to such an extent that the structure they generalize to has no foundation in imagery. In the end, symbols become utterly meaningless—and so pointless.

In the learning cycles, as recognized by Piaget, Dienes has suggested that, hand in hand with *generalization*, abstraction occupies a central place. To distinguish between these two central processes, imagine a baby learning to talk. He learns sounds such as 'chair', and from any number of isolated instances, he forms the notion of a class of objects which we label 'chair'. In finding the properties common to a wide range of stimuli, stripping percepts of all irrelevancies, Koestler sees 'obvious analogies between the abstraction of ideas—formation of concepts—and perceptual generalization—the extraction of invariant features, stripped of their accidental accompaniments, from varied situations'. He suggests that abstract thought is merely the extension, along a continuous scale, of selective processes which operate already in the rat and even lower down. We abstract—formulate classes—readily and all the time. (All too readily, sometimes, as anyone concerned with racial prejudice will know.) The next step, that of generalizing, seems to be difficult for children, possibly because of intellectual inertia. We are happier with our tried-and-tested class abstractions, which tend to be upset by new ones. Later on, the child will generalize the class of chairs by extending it to include, under the heading 'furniture', all sorts of other classes of objects.

It is tempting to see in abstraction an analogue of the cognitive process Piaget labels *assimilation* (a sort of 'ingestion' of experiences); whereas *generalization* analogizes Piaget's *accommodation* (a kind of adaptation to the environment).

Basically, abstracting is the process of forming an idea from a few or several concrete situations; generalizing is extending that abstract idea further afield—this is what a scientist is doing when he extrapolates his experimental results on a graph, or a mathematician when he asserts: 'If we take 2 things (call them xs, perhaps) and 3 things more, we *invariably* wind up with 5 things altogether.' Boldly, he dispenses even with the mystic x, writes baldly, $2 + 3 = 5$, and declares that the formula holds for anything you care to think of . . . even things *he* has never even heard of! Which is quite a feat of generalization. Finally, we can say with some certainty it is possible to abstract without generalizing but not to generalize without abstracting!

Psycho-mathematics, the science of concept formation in mathematics, was put on the intellectual map by Jean Piaget, geographer of the child's growing mind. Today one of the most prestigious centres devoted to such mental geography is Harvard's Centre for Cognitive Research; another such laboratory of 'child's-play' is Dienes' renowned centre at Sherbrooke, Canada. But the most famous is Piaget's own institute in Geneva. For fifty years Piaget has been discovering through deceptively simple experiments that children actually think in staggeringly intricate and complex patterns that, he feels, adults should learn to appreciate better than they do. His findings are at the root of the so-called 'discovery' methods of learning in primary schools. The method also draws on the ideas of Maria Montessori, the Italian educator of children, John Dewey, the psychologist, and more recently, Jerome Bruner, the Harvard psychologist.

We can say, with Piaget, 'a ready-made truth is only half a truth. The goal of education is not to increase the amount of knowledge but to create possibilities for a child to invent and discover, to create men who are capable of doing new things.' To Bruner, Piaget's conception of the growing mind 'is so compelling that even in attacking it one is inevitably influenced by it'.

In a sense, Piaget's method was anticipated by, ironically enough, Herbert Spencer with his classical *empiricism*, a philosophy that reduces everything to experience. The irony is that Piaget, while declaring that we derive our first mathematical concepts through the process of abstraction from physical objects, does not relate *everything* to experience. Like Hayek, he in effect allows that we inherit some abstract ideas as a cultural–racial heritage. Even more ironically, Russian schools share the 'experience-is-all' view, which is scarcely consistent with the central theory of dialectics of the inheritance of acquired characteristics. The horrifying Medvedev incident in 1970 showed this theory is still being actively enforced.*

Piaget distinguishes between the physical experience on which the scientist draws and logico-mathematical experience or deduction. On the one hand, logico-mathematical experience begins, as he sees it, when a child works from experience. This is paralleled in history by the Egyptians' need to measure land leading to geometry, in particular the simple case of Pythagoras' theorem about 3–4–5 triangles. On the other hand, physics involves the scientific abstraction direct from experience, such as the flattening of a bouncing ball or extension of a spring or trajectory of a ball. In physics the abstraction is direct from objects, whereas in mathematics it is 'reflective' and is not derived from objects but from what we 'do'—admittedly, mostly in the mind—to the objects. This distinction is at the heart of Russell's famous quip about mathematics being the subject 'about which one never knows what one is talking about'. Because, of course, it does not matter: mathematics is only concerned with relationships and even relationships between relationships . . . and so on *ad infinitum*, if necessary.

When a child first learns mathematics he manipulates visible, tangible objects; in time he can dispense with these aids and grasp or visualize simple problems solely with his mind. Mathematics depends ultimately on neurological co-ordination. Even physicists scorn 'experience', as for example when they subordinate what they observe about elementary particles in a giant bubble chamber to the mathematical idea of a *group*, a scientific model. Nobel-laureate Murray Gell-Mann in 1964 achieved just such an intellectual miracle, with his co-researcher, Yuval Ne'eman, when he correctly predicted the existence of the now-famous Omega-minus particle. Crudely speaking, the scientific game today is to 'bend' the facts to fit the model—or at least to select those that do! Such fruitful fitting-facts-to-models goes on in biology today and, no doubt, will one day be applied to social sciences.

Learning games and strategies

In the US, Patrick Suppes has intensively studied how children form concepts of a mathematical nature with the aid of computer-assisted systems. In Australia Dienes and M. A. Jeeves, too, isolated three main

* Zhores Medvedev, a world-renowned biogeneticist, was incarcerated in 1970 within a psychiatric ward in Russia for holding scientific views not in line with the dialectical theory. With outstanding scientific honesty and great personal courage he dared to question the Lysenkoist line—now, incidentally, totally discredited.

strategies of learning—by memory, in patterns, and by operators. Suppes has focussed his interests on two vital ingredients of concept-formation—besides the mainstays of abstraction and generalization—namely, logical deduction and *axiomatization*.

Axiomatization is the determination of a preferably small, basic set of relationships from which all others may be derived. The desire for smallness is not only practical; a compact, all-embracing formula is obviously preferable to an endless list of components that fit the formula. The well-tried rule of thumb, known as Occam's Razor, reflects the lasting utility of economy in hypothesis-making. This maxim, which was advanced by the philosopher William of Occam in the 1300s, says: 'The hypothesis with the fewest assumptions is to be preferred.' As a rule, the best models and laws are the simplest ones with the fewest elements. Indeed, it was Einstein's dying hope—never fulfilled—to produce one all-embracing theory to explain every known physical phenomenon. Give a mathematician a four-axiom system and he will inevitably see if he cannot reduce it to a three-axiom structure! Again, Gell-Mann's triumphant use of group theory to predict the Omega-minus particle rests on the mathematical structure of a group, which has only four axioms, whereas the more conventional nineteenth-century field theory, with some dozen axioms, could explain electro-magnetism but not elementary particles like the Omega-minus.

Suppes advanced a convincing case for the existence of an 'all-or-none' conditioning process in simple concept formation. Graphed, the 'all-or-none' learning pattern has a two-step appearance (see Figure 13)

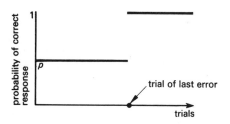

Figure 13 Learning curve for 'all-or-none' model

and leads directly to the following model, formulated in terms of a stimulus-response experiment on a child. The child-subject begins, not knowing the association which the experimenter (the teacher, perhaps) sees linking stimuli and responses. The stimuli might be logical puzzles or simple arithmetic 'sums'; the expected responses are the conventional answers. In the unconditioned state, the child will simply guess the correct response with a probability we'll call p; this guessing probability is, on this model, supposed to be independent of the number of trials performed or the preceding pattern of responses. At the outset, in Suppes' tests, the probability was some figure less than 1; a typical value was 0.4. After conditioning, the guessing probability was stepped up in one trial literally to 1, which implies absolute certainty of success in the trial. The trial results were plotted on what are called Vincent 'backward learning' curves, showing percentiles of trials rather than number of trials prior to the last error.

Now a rival 'steady build-up' learning model postulates that a child's probability of making a correct response increases linearly each time he is asked the question and shown the correct answer. This learning curve theoretically climbs steadily then flattens off at $p = 1$ (see Figure 14). The initial experiments of this project were carried out by Rachel

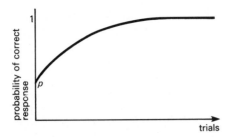

Figure 14 Learning curve for linear incremental model

Joffre in Brussels. They resembled the well-known 'oddity' tests given to monkeys. The child had to select the 'odd' shape out, by matching it with a dash-dot response. For example, eight-year-olds were shown three dash-dot patterns beneath a 'stimulus' such as two triangles and a circle in a row (see Figure 15). The child had to match one of the dash-dot patterns to the geometric shapes—in this instance, 'dash-dash-dot'. The test also investigated latencies, or reaction times.

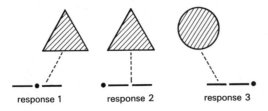

response 1   response 2   response 3

Figure 15 Stimulus display in a pattern experiment

Many teachers accept latency as a surer measure of a child's mathematical concept formation, at least, than his getting the answers correct: the time differential between the quickest and slowest pupil's response in a maths class is appreciable. (The query 'What is 6 sevens?' may elicit latencies varying from a fraction of a second to a matter of minutes.) However, in language learning, the differential may not be measurable without the aid of a stop-watch. But in the long run, these latencies mount up in the course of a year to produce quite staggering differentials in children's working-rates. Over the year a 'quick' American sixth-grader (roughly, an eleven-year-old school-child) may complete some 4000 maths problems to his 'slow' classmate's 2000. The 'quick' and the 'slow' child may be of equivalent socio-economic standing and family background and have equal IQ ratings. Teachers are rarely in a position to do much about this inconvenient fact; as well, it is generally accepted that urban children, for one reason or another, are brighter than rural children. Perhaps, like the 'city-slicker' rhesus monkeys of Jaipur, they are merely sharper and more cunning—not more intelligent.

Suppes also tested five-year-olds' flair for proving things logically. He found that their learning pattern again fitted the 'all-or-none' hypothesis. The game—for such it was to his young subjects—was to make green and red lights appearing on a console panel match a pre-set string of such lights. The game was structurally similar to mathematician-educator Paul Rosenbloom's 'golf' game developed at Minnesota, US, which is based on strings of symbols. The child had to build up a prescribed string, by using certain rules of inference. Instead of strings of red and green lights, however, we shall write strings of 0's and 1's. Four rules, $R1$, $R2$, $R3$, $R4$, were used to build on a given string S:

$$R1: S \rightarrow S00$$
$$R2: S \rightarrow S11$$
$$R3: S0 \rightarrow S$$
$$R4: S1 \rightarrow S$$

The given string S could, of course, be any string of 0's or 1's; it could even be a blank space, not to be confused with a zero. The child might be asked to build up, say, the string 101. This would be his *theorem*: to prove 101, given the string S, which was, in effect, his axiom, the irreducible, indefinable starting point for his proof. Suppes invariably started the children off on the axiom $S = 1$, shown in fact by a glowing red panel light. Here is one way (there are alternative ways, just as there are alternative proofs in geometry) of building up the string 101, that is, of proving it:

$$1 \xrightarrow{R1} 100 \xrightarrow{R3} 10 \xrightarrow{R2} 1011 \xrightarrow{R4} 101$$

By pressing buttons corresponding to each rule the child could build up the equivalent string of lights, red, green, red. Marginally, it may be noted that the coloured panel lights are, to use Jerome Bruner's term, an *iconic* form of symbolism, or what the American psychologist J. P. Guilford calls figural information (discussed later in this chapter under the heading *Symbol—servant or tyrant?*).

'How best to help children catch up?' To answer the common teachers' plaint of what to do with mid-term or mid-year arrivals to a classroom Suppes devised another 'all-or-none' experiment. An effective way to achieve this 'crash' programme of learning, in logic at least, is to teach the late-comers the formal machinery of logic first, and later to give them the interpretation of the symbols for the logical connectives, 'and' and 'or'. The results pointed to a two-stage model of learning: in the first stage children's learning proceeds more slowly than in the later stage.

Dienes has developed learning games rather than experimental routines, which employ strings of symbols linked, not by disconnected rules, but by rules conforming to a well-defined mathematical structure, such as the Klein group or the dihedral-3 group.

Transfer and autonomic behaviour

As a rule, psychologists talk about the vexed topic of transfer in two senses. First, as stimulus generalization without which, conceivably, no concept could ever be learnt. As an instance of this sense of the word, 'transfer' is involved in recognizing that two triangles and a square have the same 'oddity' as 'dot-dot-dash' in Morse code. At this point the reader might be forgiven if he gathered the impression that learning procedures are cut-and-dried processes, scientifically earmarked, lacking only the experimental psychologist's *imprimatur* to become proven fact . . . until he ponders the remarkable findings about man's autonomic system. Bernard Engel of the National Institute of Child Health and Human Development in Baltimore, US, and Joe Kamya of the Langley Porter Neuropsychiatric Institute, San Francisco, US, have found that man *is* capable of controlling certain aspects of his autonomic behaviour, normally beyond the control of all but yogi-like mystics. When a subject was rewarded for depressing his heartbeats or varying his brain waves in some predetermined manner, then, although the subject could not sense these changes, he was capable of doing so and gaining the reward—thus testifying to the possibility of controlling autonomic processes through operant conditioning. Conceivably, learning may well be controlled only in such indirect, but no less positive ways. The same remarks might eventually be shown to explain the sort of subtle ways we direct our thinking which physiologist Edward de Bono has called 'lateral thinking'.

So many people actually 'see' numbers in colours or on a map (no less an intelligence than the nineteenth-century scientist, Sir Francis Galton, had this sense to a marked degree) that it is likely that children, at least, make use of such images to form mathematical concepts. Ralph Norman Haber's experiments with children suggest that they briefly retain a photographic image of what they have seen; such after images—technically, *eidetic images*—may be another aspect of autonomic behaviour. The eidetic sense does, however, seem to fade with age, like the images themselves. But at the age when children are first

Figure 16 Alice's 'Cheshire Cat'—an eidetic image?
(*Random House*)

exposed to number concepts, the eidetic sense is markedly present in most youngsters. (This may be one reason why youngsters can be so exasperatingly good at the memory game played with cards sometimes known as 'Pelmanism'.)

What of the second sense of the word 'transfer' that psychologists recognize? This reflects lateral thinking in greater measure—that is, the transfer from one concept to another. But both positive and negative transfers are bound to happen. From a vast battery of response and latency (reaction time) tests in arithmetic, pronounced indicators of negative transfer have been found between children's ability to subtract and add. Suppes considered this specific aspect: he presented children with learning tasks based on matching corresponding sets of two and three symbols. The most obvious match is that between two identically recorded sets, with same symbols and same order, for example, $\{a, b\}$ and $\{a, b\}$. With the children's tests, the symbols used were naturally iconic—triangles, stars, shapes of common toys and so on. Suppes classified the learning of pair-associates into identity (same set members, same order of members), change of order, different membership, and different number of members. A reader familiar with the four-group will sense a family likeness which is marred, as it were, by the lack of 'closure'; the family is not a tight-knit group. Mathematically, sets are identical whatever the order their members are written in. (The order is implicitly understood to be a left-to-right reading order; indeed, it may be the very implicitness of this ordering that befogs children's appreciation of order in the mathematical sense; it is impossible to record something and abide by the Gutenberg conventions of linear print, which Marshall McLuhan has advertised so widely, without at the same time invoking order!)

Suppes wondered what happens if we 'train children' on identity of *unordered* sets, that is, the mathematical sort of set. 'Using fairly simple principles of interference confirmed in work on paired-associate learning,' he expected, and found, negative transfer in going from identity of ordered sets to sets identical in every respect. This much might be expected from the results of Dienes and Jeeves' work on 'embeddedness of concepts'. Their argument rests on the premise that generalization comes harder to a child (or adult for that matter) than abstraction, for reasons we have seen. What Suppes was testing was the ability of a child to learn about ordered sets and thence the more general notion of sets, ordered or not. Suppes found that all the negative transfer resided in the pairs of sets where the order was changed (reversed, in the case of two elements in a set). He pin-pointed little evidence of negative transfer in the subconcept of increase of the number of elements in a set—presumably because this is rather more obvious at a glance.

The Eureka effect

Suppes' view is hardly novel or surprising. Poincaré, on the other hand, pin-pointed the Eureka effect in mathematical creation. It was while walking down a street in Mont-Valérien that Poincaré made the all-vital mental leap towards the solution of the Fuchsian functions. It would not be astonishing if children thought similarly. Jacques Hadamard tells of the seemingly random role which the unconscious played in his own mathematical inventiveness:

'One phenomenon is certain and I can vouch for its absolute certainty; the sudden and immediate appearance of a solution at the very moment of sudden awakening. On being very abruptly awakened by an external noise, a solution long searched for appeared to me at once without the slightest instant of reflection on my part—the fact was remarkable enough to have struck me unforgettably—and in a quite different direction from any of those which I had previously tried to follow.'

André Marie Ampère, whose name is given to a unit of electric current, discovered solely and purely by chance the solution to a mathematical problem that had plagued him for twenty years. Karl Friedrich Gauss, assuredly one of the three greatest mathematicians who ever lived, wrote to a friend about how he proved a theorem on which he had worked unsuccessfully for four years, '. . . not by dint of painful effort but . . . as a sudden flash of light, the enigma was solved . . .' Britain's noted mathematical-physicist, P. A. M. Dirac, saw the solution to certain quantum-mechanical problems by first postulating the fantastic notion of the existence of a sea of electrons with 'bubbles' of positive electrons or positrons (later found to exist) in it; he then proceeded to prove his proposition. Undoubtedly Dirac saw the solution first; then he went on to prove his conjecture. Christopher Zeeman, a British topologist, recalls how he sliced a pages-long proof of a theorem to a few lines after years of mulling over it; again the solution came to him in a flash. There are many similarities between these intellectual feats and a child's solution of *his* maths problems. Admittedly, a child's span of attention is comparatively short, but the random element, the intellectual leap in the dark, remains constant.

In his brilliant *The Act of Creation*, Arthur Koestler refers to this leap in the dark as the 'bisociative act'. If not a random intellectual act of cognition, he sees it at least as irrational:

'In the popular imagination these men of science appear as sober ice-cold logicians, electronic brains mounted on dry sticks. But if one were shown an anthology of typical extracts from their letters and auto-biographies with no names mentioned, and then asked to guess their profession, the likeliest answer would be: a bunch of poets or musicians of a rather romantic naïve kind. The themes that reverberate through their intimate writings are: the belittling of logic and deductive reasoning (except for verification after the act); horror of the one-track mind; distrust of too much consistency ("One should carry one's theories lightly," wrote Edward Titchener); scepticism regarding all-too-conscious thinking ("It seems to me that what you call full conscious-ness is a limit case which can never be fully accomplished. It seems to me connected with the fact called the narrowness of consciousness, *Enge des Bewusstseins*," Einstein).'

In Edward de Bono's eyes, such insight may be a legacy of his 'lateral thinking'. It is probably true that what is lateral thinking for most, amounts to purely routine thinking for a professional mathematician. True too is the fact that overstraining of one's mental faculties (or emotional ones, for that matter) can seize up our thinking engine so that we can create nothing more. A certain *insouciance* is the very life-blood of creation. Or, as the French put it so well, 'Blessed are the *debonair*: for they shall inherit the world.' Certainly, the intellectual world.

Symbol—servant or tyrant?

Not even mathematicians and logicians are free from the tyranny of symbolism. A common 'trick' question asked of maths students is: 'Is this a set—($a\,b$)?' As most readers are aware, a *set* is an abstract idea of a collection of things, though it does not actually include the things themselves. Mathematicians have come to write the set of two things to which we have given the label a and b as $\{a, b\}$. We said it was a trick question: the dutiful student is expected to write curly braces instead of parentheses and, of course, on no account must he omit the vital comma (which merely separates the items within the set). Such trick questions are a favourite of programmed learning texts or tapes which often give the impression of being specially prepared with the intention of tripping up the unwary. Of course, to return to our 'sets' example, it does not matter a jot how we write a set, as long as we convey our meaning and can operate with our symbolism effectively.

Since mathematicians deal almost entirely in symbols, it is best to be clear about the link between symbols and the concepts they symbolize. C. K. Ogden and I. A. Richards in 1930 were among the first to point out the threefold distinction of *referent* (the thing perceived), *symbol* (speech sign of the thing), and *reference* (the thought about the thing). Mathematicians are almost exclusively concerned with the reference, the thought, rather than with the referent, the thing itself. So that the reference *refers to* the thing, and the symbol *stands for* the thing, though it would be more appropriate to talk of a *sign* rather than a symbol. Symbol is strictly a more general term embracing such things as flags, coats of arms, trade-marks and so on. Sign is used more restrictively of language units—visual and auditory. The important point is that a sign can either symbolize the thought or stand for the thing. So much for the semanticist's view, illustrated in Figure 17 by

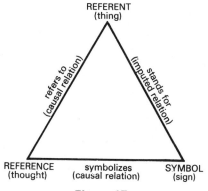

Figure 17

a model presented in 1930 by Ogden and Richards. The mathematician, on the other hand, employs signs to stand for a bunch of signs, each of which refers not to a thing but to an operation. He gaily uses the same sign to stand for many different things or operations.

Of course, a mathematician will use symbol *x* as shorthand, though not quite in the way we call Shakespeare's most famous play by the shorthand symbol, *Hamlet*. To recall the play we do not need to recite all of it. But the mathematician is apt to designate a symbol for a relation between two such things as *Hamlet* and *Two Gentlemen of Verona*. He might invent a symbol ⇝ to remind him that one play was written before the other, and he could then write: *V* ⇝ *H*. This shorthand recalls that *Two Gentlemen of Verona* was written before *Hamlet*.

J. P. Guilford categorizes content and information into figural and symbolic information. He sees symbolic information as being 'in the form of signs, materials, the elements having no significance in themselves, such as letters, numbers, musical notations, and other "code elements" '. He notes that oral speech has its code elements in the form of about forty phonemes, basic sounds from which all spoken words can be formed. Figural information he defines as being 'in concrete form, as perceived or as recalled in the form of images'. Different sense modalities may be involved—visual (as with the red and green lights on the panel), auditory, or kinaesthetic (the sense of muscular movement). The first kind of figural information a child encounters are the letters between which he must learn to discriminate if he is to learn to read. He must learn to recognize each letter despite such transformations as it undergoes when viewed from different angles, rotated, changed in size, and distorted in innumerable ways as with different type faces. With all this a quite average child can cope—and more. The ease with which a child, or anyone, can read black letters on a white background (as on this page) or the reverse (as on a blackboard) is due to the way our eyes see outlines rather than spaces inside boundaries. (See the section *A Model of the Brain* in Chapter 5.)

The ability to read handwriting, of course, depends on the high redundancy of language. In English it is estimated at about 70%. This high figure reflects the basic argument of how we should teach children to read. Rudolph Flesch has drawn attention to the absurdly high redundancy in many American first readers, compared to the more rational approach favoured in most British texts. Redundancy is calculated as follows. Let us take, say, a three-word sentence: 'See Johnny play.' There are three words, of which none is repeated. The redundancy ratio is the total number of words (3) less the number of different words (3) divided by the total number of words—that is, 0%. A normal redundancy ratio in a five-word sentence is 1 in 4 or 25%—

3 different words, with 1 repeated. In a typical American reader, of the type criticized by Flesch, there were 239 running words, of which 47 were different; that is, 192 were repeated. So the redundancy ratio was 192/239 or 80%—far too high, and far too boring to read.

Redundancy can be useful, however. Every political speaker soon learns the need for repetition. Shakespeare often used messenger and choruses to repeat the plot, sometimes at least three times, because he probably knew that at any given moment only about 25% of the audience would be listening attentively, that any one playgoer would only absorb a quarter of the whole play in one performance. Shorthand makes use of redundancy in language by assigning a special sign, a logogram, to a frequently used word or phrase. Too much redundancy can actually inhibit the transference of information. A new-minted phrase or word can be used to death and become a lifeless, meaningless cliché.

A child's learning to read reflects on a small scale what the mathematician has to do when learning a new concept. Piaget outlined a three-stage development for children's learning and intellectual development: the play stage from cradle to seven years old, the concrete operational stage from seven to twelve years, and finally at puberty and thereafter the formal operational stage. Some authorities consider that the same learning cycle may be detected in the learning of any new concept, a case of a micro-cycle mimicking a macro-cycle, as it were (though Piaget would not, apparently, agree with this extension of his original idea!). Once the child has learnt to distinguish various marks on paper, he must translate these visual units into auditory units—that is, graphemes into phonemes. The bugbear of learning English (for a foreigner) is that the one grapheme (roughly a syllable) is encoded with many phonemes, as a rule. The relation is, as mathematicians would say, a 'one-many' mapping, with which mathematicians have as little truck as they can. It is to avoid this one-many situation that modified alphabets, such as the initial teaching alphabet, are given to children who are learning to read. Remarkably, children later make the transition to the conventional alphabet without much trouble. Reading and writing in whole syllables takes on the form of a coding and decoding process. And so it is with mathematical symbolism.

First a child, or learner of any age for that matter, can be presented with iconic information as figural units—triangles, homely objects and so forth. Then the same information can be coded in sign form *when the concepts have been learnt*. Often this is done when they haven't! And therein begins a maths block.

Gaston Viaud sees a concept as

'a generalized and abstract symbol; it is the sum of all our knowledge of a particular class of objects. Being abstract, it contains, or rather allows us to recall, the characteristic properties of the class of objects that it symbolizes, and that distinguish that class from all others. For example the concept "dog" is the sum total of the properties that distinguish the dog from all other animals: it is a mammal, it barks, it runs quickly, it hunts, it guards sheep or houses—and so on. Being a symbol, it refers also to all members of a class, for example to all animals which answer to the description "dog". Thus, my concept "dog" includes all I know about dogs . . .'

For a child the trouble with much mathematical symbolism is that he knows how to write 'dog' (in mathematical terms) but has never seen or patted a mathematical 'dog'. So he does not know what he is writing about (symbolizing). Present even a mathematician with a system of symbols but do not tell him what the symbols represent; then give him the rules for playing with the symbols and he will probably be stuck hard and fast intellectually. This is precisely how a child feels who has been taught a mass of squiggly symbols before being properly taught what they represent. It is rather like reading Keynes on economics (see Chapter 6) without knowing what money is! The recognition of signs extends to animals: Koestler muses on how a dog recognizes another dog across the street—at that distance, a walking symbol.*

It seems the mathematician has, at some time, to divest a symbol of its contextual meaning if he is to make the widest use of it. At Harvard's Centre for Cognitive Studies, Dienes observed children generating and, indeed, playing around with their own do-it-yourself symbols. He found that the introduction of symbols too early in the game could actually impede concept formation, though sometimes the introduction could lead to creative thinking. Dienes regards a symbol—or in Guilford's terminology, a sign—as being nothing more than a chalk mark or trace of printer's ink. Dienes also advances the notion that it is impossible to make the bricks of complex abstract concepts without the straws of symbolism. 'The mathematician's job', he posits, 'is the *making* and the sorting of heaps'—classes and classes of classes heaped together by some defined relationship. Of course, signs (or symbols) must not be confused with the things they symbolize. By no stretch of the imagination can the word 'food' be likened to the thing it stands for.

* Probably an olfactory or auditory symbol.

Heinz von Foerster—one of the new breed of neurophysiologists who, like mathematician-topologist Zeeman, are trying to bridge the gap between the micro-concepts of the way tiny synapses in the brain function and the macro-concepts of how we behave as large animal entities—suggested the following apt analogy. On landing from an air flight we might be asked about the food served by the airline. We answer, very properly: cold meats, salad, bread and butter, sweet (undefined), and red wine. Now nobody in their senses expects us to produce the real foods in response to the query. Nor, be it noted, did the stewardess tell us what the courses of the menu were, and we never read a menu. We just ate the items! Our experience had been transformed into utterances—that is, symbolic representations of these experiences.

To E. H. Gombrich, the hallmark of a symbol is the feeling that it is more than a sign (the distinction was drawn earlier in this section): it affords us glimpses of vistas of meaning beyond the reach of convention or logic. 'The longing for context in cerebral space-time', he writes, 'represents the leap across Euclidean co-ordinates and a search for meaning beyond the logic of survival.' And he invokes Pip's words in Dickens' novel, *Great Expectations*, to illustrate his point.

'As I never saw my father or my mother and never saw any likeness of either of them (for their days were long before the days of photographs), my first fancies regarding what they were like, were unreasonably derived from their tombstones. The shape of the letters on my father's gave me an odd idea, that he was a square, stout, dark man, with curly black hair . . .'

Like Pip, man can never reconcile himself to sheep-like acceptance of a human condition based only on the use of signs, in what Roland Fischer calls 'physical survival space-time' (discussed later in this chapter under the heading *Space-time*). The reader may object and say that what Pip couldn't stomach was that it was his parents' names chiselled in dead letters that he was trying to decode. An effigy would have been vastly more representational and comforting. But, in fact, an iconic sign—for such an effigy is, like the Egyptian hieroglyph (a cartoon sketch of, say, an eagle for the real thing)—is an inefficient means of coding information. The word 'iconic' means like an icon or picture; it has to do with images, graven or otherwise. The letters e-a-g-l-e are vastly more effective in conveying this information than a hieroglyph or other iconic symbol; the collective symbol recalls the abstraction of 'eagle-ness', traced from many eagles yet recalling no one *particular* eagle.

To a mathematician, symbols are catalysts of thought. The very act of putting pencil to paper stimulates the flow of his ideas. Some mathematicians never become reconciled, however, to symbols learnt after childhood, it seems: some have to translate Greek symbols to their Latin equivalents before they can 'read'—that is, decode—the mathematical message. This may be a subtle form of mnemonic for if we cannot say a word it is well-nigh impossible to remember it. 'Words,' said Bergson, 'are the carriers of memory.' Human memory becomes organized by virtue of speech alone. Since a good memory is a 'must' for effective operation in mathematics, the role which symbolism plays is of paramount importance.

The symbol reminds the mathematician in one glance of the mathematical structure he is trying to handle. Moreover, symbols can represent whole chunks of mathematical operations at a time; so, by using them the mathematician is able to skip a great number of steps; in fact, he can allow the symbol, as Dienes sees it, 'to do part of his thinking for him'. The symbolic mechanisms in this way act as primitive computers, accelerating the thought process, saving mental energy for those parts of the process which are not mechanical and which still have to be put together for the first time. Complete theoretical as well as practical mastery of the techniques is essential for this means of acceleration to function, as otherwise it will not be possible to interpret the results reached at the end of a mechanical stretch of the mathematical road.

Hadamard and Wertheimer pin-point another use of symbolism: as a shorthand common to many different mathematical structures. The ability to perceive a common structure in two apparently disparate mathematical situations is often a vital prerequisite to really original thinking. The symbols, of course, do not embody this 'same structure' feature, which mathematicians know as *isomorphism*: they merely enable the mathematician to discern the structure at a glance. To take a crude example, the letters *rgr*, where *r* stands for red, and *g* for green, have an obvious, easily assimilated pattern; however, the pattern of giant posters of the same colours arranged side by side might conceivably elude the eye.

For good reason, our discussion of symbolism has ignored the effect of ambiguous shapes, such as the famous Necker frame cube (see Figure 18). Obviously such a symbol, difficult enough to draw, would be decidedly unsuitable as a mathematical symbol. But colour can be used constructively and meaningfully as mathematician Arthur Cayley

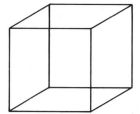

Figure 18 Necker frame cube

discovered. On the whole, auditory and kinaesthetic symbolism is ruled out for mathematical purposes on the grounds of being time-dependent: it is not possible to look back over such symbols as is the case with printed signs.

A mathematician's philosophical outlook may well colour his attitude to symbolism; this attitude will probably be governed by his reaction to the formalist-intuitionist battle that raged at the turn of the century. Russell, and other formalists, regarded symbolism as the essence of a great number of mathematical situations, though symbols are devoid of content in themselves. Most teaching today favours the quick learner of mathematical processes (algorithms) or the logical thinker. But the constructivist, whose life-blood is a multiplicity of embodiments, hardly gets a look-in. Dienes writes:

This may be why constructivist* mathematicians are so few and far between. If the constructive processes of building classes, particularly isomorphic ones, were allowed more space in our mathematics syllabuses, we should probably find the balance much more even between the two types of thinkers amongst our mathematicians.'

Other geometries, other spaces

'Our school geometry', writes Charles Proteus Steinmetz, one of America's great electrical geniuses and no mean mathematician either, 'is built on a set of axioms which was selected by Euclid 2000 years ago . . .

'Undoubtedly, Euclid was led by experience when selecting the axioms which he chose, and therefore the theorems of Euclidean mathematics have been in good agreement with physical experience.

* The Constructivist, or Intuitionist, school of mathematics was founded by the Dutch mathematician Luitzen E. J. Brouwer in 1912. For a non-technical account see Michael Holt's *Mathematics in Art*, Chapter 6.

This, however, is no part of mathematics.' The nineteenth century saw the creation of wild and variegated hot-house geometries, apparently without root in the physical world. These intellectual blooms were grown by such great mathematicians as the Germans, Gauss and Georg Riemann, the Russian, Nikolas Ivanovitch Lobachevski, and the Hungarian, Johann Bolyai. So, Steinmetz proposes, 'when finally, in the relativity theory, physics advanced beyond the range of Euclidean geometry, the mathematics of the new space characteristics were fully developed'.

Euclid's axioms, the reader may recall, were statements such as 'Two points determine a straight line' (or more probably indelibly printed on the mind, 'The shortest distance between two points is a straight line') and 'Two straight lines with two points in common must have all points in common.' But the most troublesome and most contentious was the notorious Fifth Postulate, the so-called 'parallel axiom': 'Through a point outside a given straight line one, and only one, parallel line can be drawn.' Euclid defined a parallel as a line in the same plane as the given straight line which, however far it stretches, never cuts the given line.

For nearly 2000 years the axiom stood inviolate but suspect; nobody, not even Euclid, had proved it yet many suspected its legitimacy. Finally, the nineteenth-century mathematicians gave up the struggle to find a proof and adopted another ploy. They said, in effect:

'Just suppose the parallel axiom is not an axiom but a law that can be derived from other, more fundamental axioms [such as the ones already stated]. Well, then, choose a different parallel axiom—for instance, perhaps more than one parallel exists, or none. Now if our new-style axiom leads us into contradictions with the other axioms, the parallel law is not a respectable axiom. On the other hand, if it does not lead to contradictions, we shall get a self-consistent system as watertight as ever Euclid's was!'

In the event, the mathematicians did develop intellectually watertight systems and the parallel axiom was vindicated as being able to stand comparison in its own right with all the other axioms. Euclid's geometry is founded on the axiom of one parallel. But if we posit other axioms, we get other geometries relevant to other kinds of space. For example, more than one parallel gives the 'hyperbolic' geometry, and no parallel gives 'elliptic' geometry. The geometry of several parallels applies to space in a gravitational field; 'no parallel' geometry applies to a centrifugal field (as in a washing machine spin drier). In the next section we look at ways we learn about the space around us and formulate our findings in a geometry.

Space and shapes

Certain aspects of geometry teaching in primary school bear a strong resemblance to problems of reading. Just as there are reading invariants —this sentence can be read if printed in all sorts of type faces and sizes, colours, and on such various media as newspaper, book, TV or film screen, or as smoke-writing in the sky—so there are spatial (geometric) invariants. Once children were brought up on Euclidean geometry, where the properties of shapes, such as congruence, similarity and parallelism, were studied as if the shapes were drawn on a rigid plane— hence the name, *plane* geometry. The fact is, such a plane is no more than a physical fiction, an abstraction of the mind. Maybe ancient geometers abstracted the idea of a plane when drawing in the sand or through observing the mirror-like surface of water in a jug. Under- standably, ancient thinkers were slow to follow in the footsteps of radical thinkers like Ptolemy who thought the world to be round. Lakes and other small water surfaces lent no credence to the then held belief that the sea's horizon was curved; alert mariners observed that this was so as their ships 'sank into the distance'.

Next, the orientation of the plane—here, our drawing board on which we create our mathematics—has, as a rule, to be ignored in most classrooms. As Piaget noted, children before the onset of the second stage in his developmental scheme (when they are in the 'sensori-motor' stage, as he calls it) experience intense difficulty in grasping the con- stancy of water level: they do not believe that the water surface in a jar of water does not tip with the jar when tilted. One of the authors who performed this and other experiments of Piaget's found that although Piaget had unerringly put his finger on the conceptual nub, children's responses might be positive or negative according to Piagetian criteria. In our view Piaget may be said to have isolated the essence of the prob- lems that children encounter if not always the exact polarization in the way they react to them. It might be wondered why so few of Piaget's 'conceptual' experiments on children have been repeated and confirmed or refuted, particularly since his own experiments were conducted with a curiously small section of not especially typical or average Swiss children; for, continual and massive verification is part of the scientific game. According to Karl Popper's canon—that a scientific theory to be worthy of its name must be prepared to stick its neck out—the burden of validity rests on its resistance to disproof.

The point is simply this: first, children are expected to abstract the concept of a plane at the outset of their education and, in some instances, to ignore its orientation. Over the issue of how we should teach geometry, looms the apocalyptic figure of the great nineteenth-century German mathematician, Felix Klein. Klein in his justly famous Erlanger Programme advanced a new line in the teaching of geometry when he epitomized it as the study of the invariance of rigid shapes even under motion. His approach, invoking the idea of 'invariance', prompts the question: 'What stays the same in a flat shape if anything can happen to it short of its being torn or punctured?' The answer led to the creation of the branch of mathematics known as *topology*.

A thought-experiment may clarify the distinction between the two geometries, Euclid's and topology. Think of a triangle, of any sort, drawn on a sheet of plane paper. Rotate the sheet, reflect it in a mirror, shift it around, and the Euclidean properties of the triangles remain blithely unchanged. The triangle does, however, go through a number of rigid motions and mathematicians call such changes 'transformations'. The transformations of one triangle comprise a special kind of set, known as a 'group'.

What happens when we put a square through certain topological transformations? (A square is better than a triangle for demonstrating the qualities we are looking for.) In the first place the sheet can become bent 'out of true'—and the square with it. Worse, the sheet can be crumpled into a ball—and even hurled into the wastepaper basket—a dramatic example of motion but, this time, not in a plane. Then, we could draw the square on a rubber sheet and stretch, twist and shrink it; an effective demonstration can be made by drawing a square on a rubber balloon and then inflating the balloon. Yet through all these violent transformations, we can descry in the buffeted square certain constancies, certain invariances. By labelling the square's corners *A*, *B*, *C*, *D*, we can discern that through all such topological transformations the order in which the corners come remains the same: *A* is connected to *B*, *B* to *C*, *C* to *D* and *D* to *A*; but *A* never becomes connected to *C* nor *B* to *D*, and *vice versa*. 'Connectivity' becomes the prime invariant in topology (see Chapter 8).

Both Suppes and Dienes noted that children when presented with a square shape, drawn square to the frame of the page or blackboard, instantly recognized it as such; but, when faced with the otherwise identical square twisted round so that it was no longer square to the frame, they saw it as a 'diamond'—different shape altogether. The authors, too, have found nine-year-olds who, though they instantly recognize a triangle at sight when its base is horizontal, cannot name the shape in other orientations! (We found that asking a child his name before, during and after doing a cartwheel drove home the constancy of triangularity both non-verbally and amusingly.)

For the child, a shape is not invariant throughout a group of rotations. Flesch in *Why Johnnie Can't Read* recounts how a child presented with the word 'John' at the top of the page is mystified by its reappearance lower down. The fact is, Flesch explains, the first 'John' has a different background of symbols to the one lower down. The child must make a conscious effort to ignore such background 'noise'. Fortunately, as in mathematics learning, children seem to be programmed to do this without being specifically taught to do so. The price we pay for such a lackadaisical approach is that of having to repeat lessons.

With the squares, in Dienes' words: 'Children have not, on the whole, had sufficient experiences of squares in varied positions . . . The class formed by them is of "squares with sides parallel to the frame" instead of the class "squares". The concept of squareness has not been fully formed at this stage.' In this observation rests the reason for teaching children topology before Euclidean geometry: it comes more naturally to them, if not to the teachers, conditioned to an earlier intellectual regime. Suppes tested the effects of rotation and size-change on a child's recognition of triangles, squares, and regular pentagons. (Figure 19

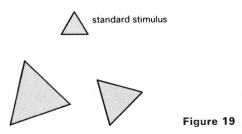

standard stimulus

Figure 19

shows a typical stimulus display in an experiment on similarity of geometric forms.) Californian five-year-olds, he found, chose the shape, whatever its size, that had been turned round the least—a tendency that decreased markedly with older children. Suppes' result seemed to chime with the findings of Stuart Sutherland's survey of experiments, *Shape Discrimination by Animals* (1961).

Animals, this monograph shows, transfer badly when a pattern is rotated, though they are able to take in their stride patterns blown up or reduced—much as Suppes' five-year-olds could. For instance, J. Z. Young's work with octopuses showed they can discriminate between 'upright' and 'horizontal' rectangles. Mammals have the power of shape discrimination to a surprising degree. American psychologist Ronald H. Forgus' more recent experiments on rats showed they can recognize triangles. The rats gave every evidence of being able to abstract the 'triangle-ness' of a triangle: they could see a partial triangle every bit as well as a whole one (see Figure 20). The rats, at the expectation of

Figure 20

being fed, had learnt to distinguish between a triangle and a circle. In the test run they were called on to discriminate between a circle and a partial triangle—either the 'all-sides' or the 'all-angles' triangle. The rats discriminated far better with the 'angles' than the 'sides'. The 'angles' triangle contains more information, Forgus reasoned, than the 'sides' triangle, in so far as an angle has two sides embedded in it. (This is, of course, true provided the triangle's size is not alterable.) As a result the rats could recognize the 'triangle' in the 'angles' more readily than in the 'sides' figure and so could more easily discriminate between the 'angles' and the circle. We might suppose that children, too, must make similar intuitive leaps all on their own, otherwise they would never 'see' a triangle in imperfectly rendered drawings on blackboard or paper.

As so often happens in science, Suppes' neat and tidy theory was smashed to smithereens by later experiments he carried out in Ghana— to trace the same effect. Ghanaian youngsters, particularly those from illiterate families, were not in the least confused by turning shapes. This set-back prompted further research with even younger Californians who, predictably perhaps, at three and four years of age took rotation equally in their stride. Suppes advances a convincing explanation for the behaviour of the American five-year-olds in particular: it is, he notes, at just this age when children are first taught to read. They ignore changes in size of letters—skill in distinguishing size has to be learnt in mathematics especially with indices and suffixes such as $x_i{}^2$—but become sensitive to small changes in the placing of letters. A child's alphabet is always shown in a standard linear orientation. It is never rotated for the good reason that 'right-way round-ness' of public signs is often highly informational. Brian Knight, a mathematician known by one of the authors, taught himself at a mature age to read upside down. But there was a snag: when he read a 'Halt' sign painted on a road he became confused as to which line of traffic it referred to—his or the opposing line!

Finally, such work on rotation of shapes might be seen simply as a facet of a personal style of thinking or 'cognitive style', which we look at later in this chapter in the section *Perception and learning*.

Space-time

What do we mean by space or space-time? Roland Fischer of Ohio State University postulates that the fabric of reality is woven from three different space-times (which he sees as aspects of various mathematical disciplines): cerebral, sensory and physical space-time. Putting these space-time models into perspective, we can view them as part of a cybernetic model of man. He looks at a flower, say. The information he gathers about the flower, what he perceives in fact, in this model, flows from the flower existing in physical space-time; this information from the man's environment flows next into a filtering device, his concept-forming mechanism of cerebral space-time; the final outcome of this flow is the image output—a mental picture of the flower in sensory space-time.

In cerebral space-time, we normally think and traditionally form mathematical concepts; for instance, in physiological terms it is here we do our perceiving. And here the laws of symmetry, identity and the excluded middle (a cornerstone of logic, as we see in the section *Tweedledee logic* in Chapter 3) simply do not hold; time can run backwards as in American physicist Feynman's sub-atomic world. (It was appropriate that time-reversal should be an American construct, with their national obsession for saving time.) And here the material of the memory is imageless, non-verbal and anything can stand for anything else. In this 'space' patients under LSD experience 'visual imagery'—

probably not unlike the concept-forming daydreams of a creative mathematician.

The geometry of sensory space-time, which for Fischer means *visual* space-time, is Riemannian (after the great nineteenth-century mathematician). This geometry has something in common with the picture world of classical Chinese artists where distant objects may be depicted far larger in apparent size than a relatively near object—even a relatively important one. Trained observers have described the perception of a rotating oblong or square as changes to an elastic trapezoid.

'Most of our waking time is spent in Euclidean, physical space-time,' writes Fischer of our down-to-earth, engineering or survival space-time; 'it is constructed from trial-and-error learning in such a way that the brain events evoked by sensory information from the retina of the eye [our looking at a flower, for instance] are interpreted so that they give a picture of the "external" world . . .' And we sense this 'external' world by touch, movement, attending and learning.

Even this brief run-down on some aspects of recent work on perception shows that the models of perception rely on mathematical constructs. Most astonishing is our innate reliance on Riemannian geometry, in which the conventional notions of parallel lines not meeting and space being flat are no longer upheld; it is from Riemannian geometry that Einstein's general theory of relativity is descended.

Visual perception

Tests on perception of triangles and circles have been made on adults blind from birth but whose sight has been restored. In the 1930s, M. von Senden collected case histories of sixty-six such cases. He discovered that all the adults exhibited an immediate appreciation of topological properties, much as has been found among children. But he also found in them an inability to detect the difference between a triangle and a circle. Some of his patients developed a technique of running the eye round the boundary and of counting the corners. (Chapter 4 explains why we see the outlines of shapes rather than their 'filled-in mass' for purely physical reasons.) But it took his patients months before they could differentiate between the shapes spontaneously. Experiments on the *saccadic* movements of the human eye (the continuous jerky shifts of glance) show that normal people can see an image flashed on a screen too quickly for the eye to scan the image.

Zeeman (whose model of the brain is described in Chapter 5) explains the difficulty experienced by previously blind patients.

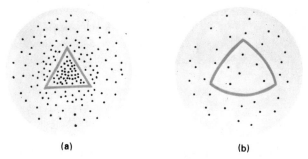

(a)   (b)

Figure 21 Sensors in the human eye bunched together at the centre in (a) and their distribution, as 'seen' by the brain in (b)

The sensors (rods and cones) in every human's eye are spread over the retina more thickly in the middle than at the periphery (see Figure 21(a)). But the sensors are linked, as by telephone wires, one-to-one with the brain cells in the visual lobes of the brain. However, the brain cannot 'know' that the receptors are bunched together and can only 'know' which sensors are close together. The brain thus 'sees' the sensors as evenly spread out. The effect is much like that produced when we blow up a balloon on which is drawn a straight-sided triangle; when the balloon is inflated, the triangle has curved sides and any dots in it will be more evenly spaced out (see Figure 21(b)).

This evidence implies that we *all* have to learn to see straight lines as such; that we do not perceive distances to be related in a Euclidean way; that is, we do not perceive a *metric*. A metric is, if you like, a reasonable way of regarding distance, whereby the shortest distance between two points is a straight line. Lack of metric throws this reasonable assumption out of the window. Because of the lack of metric in the way we perceive the world, we are able to adjust to the absurdly large-sized figures on a cinema screen—the 'constancy' phenomenon—and can even accommodate to a curved screen. Another instance of the constancy effect is the way we view ellipses in a drawing as circles. Psychologist R. H. Thouless has had people compare a circle seen obliquely with several ellipses of varying *eccentricity* (a measure of how elongated they are); surprisingly, subjects selected *different* ellipses in choosing a match for the circle. This suggests that we each have our own constancy equation.

The evidence recalls Richard Gregory's later work on the importance of being able to touch something before being able to see it. Gregory, too, experimented on someone blind from birth who had his sight restored. As soon as the bandages were removed, the man was shown a bus; the only parts he could draw were the parts he had previously touched: the stairs, handrail, and footboard. Even after many months with sight, he never drew the front and bonnet of the bus in his drawings of it. He could not see it because he had never touched it. Psychiatrist J. R. Smythies suggests that this experiment is crucial in disproving Skinnerian ideas. Other work confirms the link between the tactile and the visual senses in the sense advocated by Marshall McLuhan. From the mathematical aspect, most of us probably require far more tactile, structured experiences to gain insight than a natural mathematician.

Perception and learning

How much we rely on the visual sense for learning mathematics may be gauged by the content of the major pioneer maths projects. The Madison Project in the US, the Nuffield Project in England, the Sherbrooke Project in Canada, for example, all aim to teach children maths through all their senses—or, at any rate, through their hands as much as their eyes. Tamás Varga is promoting similar teaching, as yet against great odds, in his native Hungary. Probably the most non-verbal and purely tactile approach is being advanced by the New Guinea Cognitive Research Unit where, under Dienes' inspiration, Papuans are being educated in 'modern' mathematics to assist in their sudden transition from a Stone Age culture into the twentieth century.

There is, of course, a sound biological reason for teaching maths by visual clues. Walter S. Boernstein, a distinguished researcher at New York University, and formerly, until evicted by the Nazis, a worker in Germany, has clarified the connection between perception and optic imagery. Perception in all forms, he says, is an adaptive tool, obviously most highly developed in man. Out of the optic sense organ has evolved man's optic imagery, which underlies practically all mathematical creation. But as the Cambridge philosopher, A. N. Whitehead, once said: 'So far as reality is concerned, all our sense perceptions are in the same boat, and must be treated on the same principle . . .' The Oxford philosopher, A. J. Ayer, sees visual information merely as playing the predominant role in man's construction of his world-picture; but, unlike Whitehead, he thinks it is different from other types of sensory information. J. Von Kries, the leading physiologist in Germany in the

1920s, regards certain problems of psychology and logic 'as a direct outcome of the problems of sensory physiology'—that is, of perception.

Despite these varying views, most perceptual experiments are given over to vision; certainly *gestalt* psychology relies almost totally on vision. There is good reason for this, as Boernstein points out. Perception in man, at least, has a hierarchical organization—like mathematics itself. On the lowest level there is a genetic perception of 'brightness', the most elementary form. Above this there comes an appreciation of figural writing (which leads to identity) and finally social perception. This tallies nicely with Guilford's model of intelligence. Guilford splits the content of intellectual information into four categories: figural, at the most basic level, through symbolic (mathematical symbols, musical notation, letters of the alphabet) and semantic (meaning of words) to behavioural or social at the top level. These are the raw materials on which the intellect has to work. In particular, he describes figural information in this way:

'In the visual area, we encounter such properties as colours, shape, texture, size, continuity, and dimensionality. Shapes may be geometric or quasi-geometric, rectilinear or curvilinear and contours may be rough or smooth with an almost infinite variety possible, realistic and non-realistic.'

The senses are arranged as rungs on a ladder climbing to the topmost visual ring. As M. Merleau-Ponty put it: 'Chaque sens a son monde.' To each sense its own universe. Now the earliest of the three stages in a child's cognitive development is, in Piaget's phrase, the sensori-motor phase (the play stage). An understanding of the need for a child to play with things is central to much of the modernist's approach to teaching mathematics or other subjects. The need is echoed in McLuhan's view of communication today as 'audio-tactile'. Better still, it recalls Fischer's apophthegm: 'The proof of the sensori pudding is in the motor eating.' It could be the slogan for children's early education in general and for the teaching of geometry in particular.

The truth of Fischer's saying was compellingly borne out in 1963 by R. Held and A. Hein's kitten carousel experiment. Kittens were paired in the carousel (see Figure 22) so that one kitten was passive and the other active. The active kitten wheeled his passive partner around in the light for one hour while both spent the rest of the day in total darkness. As a result, both kittens received the same amount of visual stimulation. But the passive kitten, perhaps not unexpectedly, showed marked

Figure 22 The 'kitten carousel' experiment

inabilities in visual tasks which require depth perception and visuo-motor co-ordination. The active kitten showed no such impairment.

As well as light, gravity plays a leading role in shaping the sense modalities of our perception. The perception of the two phenomena—light and gravity—are combined to shape our personal mode of 'seeing' which American researchers Herman A. Witkin and Philip K. Oltman call *cognitive style*. In a series of brilliant experiments they have found human beings to show various self-consistent modes of perceptual and intellectual activity—their cognitive styles, in fact. At one end of the style spectrum are the field-dependent types, and at the other end the field-independent people. Field-dependent people allow the organization of the field as a whole to dominate their perception of its parts. In one series of experiments subjects were tilted in a chair, which by some mechanical contraption made the room they were sitting in (actually a large-sized box) tilt with the chair. Even when tilted as much as 35° from the upright, field-dependents instructed to keep their eyes open reported 'that they could feel no pressure on either side of the body' and answered yes to the question 'Is this the way you sit when you eat your dinner?' The crux was, they had kept their eyes open: the moment they shut their eyes they readily adjusted their body to the tilt. Field-independents, on the other hand, were well able to keep their bodies upright even when the visual field (the 'box' room on gears) was tilted.

This clear evidence for such varying 'cognitive styles' suggests one reason for the mind-block that freezes children when matching figures to the frame of reference. It might be interesting to give children only trapezoidal paper on which to draw regular shapes. Without a 'Cartesian frame' to draw on, children might then experience less field-dependence in recognizing shapes. Since a large and significant proportion of time in a maths lesson is spent copying, cognitive styles may have a distinct bearing on the way we learn maths.

Results showed that

'field-independent perceivers have a relatively articulated body concept . . . The body outline [they draw] is typically drawn in realistic proportions; body parts are included and represented realistically. Sexual characteristics are indicated and often an attempt is made at role representation, suggesting a sense of the uses to which the body may be put. In contrast, in figure drawings of field-dependent persons we find very little detail and unrealistic representation of the parts.'

This effect seems to reflect on a person's ability to visualize and handle space. Indeed, when we relate cognitive style to seeing figures embedded in complex patterns the dichotomy between styles still remains.

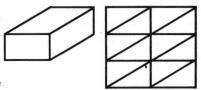

Figure 23 Thiéry's figure

Figure 23 shows a typical embedded-figure test item, recalling the well-known Thiéry's figure, which one of the authors has described elsewhere.* For field-independent people, the simple box figure on the left virtually 'pops out' of the complex design on the right; for other subjects the simple figure remains perceptually 'fused' with its organized background. We can imagine, too, that some children simply cannot *see* certain geometrical figures drawn on the blackboard; and by so much less then can they see patterns among symbols in equations in algebra, where pattern spotting is a prerequisite for success.

Success in mathematics learning, it seems, depends to an astonishing extent on learned skills and innate abilities, the marshalling and control of which are left almost totally to chance. Some improvement in mathematics learning, then, may emerge from a sound appreciation of such perceptual skills.

Non-Royal & Ancient Golf

Emil J. Post's word games of the 1920s uncannily foreshadowed present trends in language research, that is, structural linguistics. Central to the

* Holt, *Mathematics in Art*, Chapter 3.

structure is the string, or *concatenation* as linguistic experts call it. The process of stringing words together to make a concatenation is, to Paul Rosenbloom, the American mathematician, like a game of building a sequence of non-linguistic symbols—asterisks, tally-marks, or numerals. He views the acquisition of a language, too, as a sort of game, which Roger Brown dubs 'the Original Word Game'.

'Changing word order in English,' writes Brown, 'so as to reverse subject–object relations, has a consistent meaning and so directs us to the non-linguistic changes that may be expected to accompany this juggling of linguistic sequences.' A child learning English, for example, hears the word 'open', and later the utterance 'open*ed*'; the ending '-ed' to the sound identifies it mainly as a verb (though it could be an epithet as well) and suggests to the child that something in the real, non-linguistic world of actuality has changed; and the change elicits the need for so inflecting the verb. In game terms this means we can categorize as a class words like 'open-ed', 'jump-ed', and even 'carry-ed' (if we cannot spell). It is, however, curious that most of the words children first learn to read are not of this class of weak verbs, but are strong verbs—that is, they alter their stem for the past tense—such as 'ran' (not runned), 'break', 'swam', and so on.* More formally, we can write this class of weak verbs as 'Aed', where A is the immutable verb stem, or we could have a class 'Aing' or 'As'. Instead of playing with suffixes, we can find stems to fit onto prefixes such as con-, de-, in-, re- and per-. (The stem 'sist', for instance, fits here.)

To transpose our theme into a mathematical key we can pursue this idea to its logical, if improbable end. We allow ourselves two rules: the S rule and the D rule. The S rule is A becomes $A1$ (where A is the given string); this is expressed more succinctly, along with the D rule, below.

$$S \text{ rule:} \quad A \rightarrow A1$$
$$D \text{ rule:} \quad A1 \rightarrow A1 \, A1 \quad \text{or} \quad A \rightarrow A \, A$$

where A is some string or other. As in the Suppes' logic game we may always designate A as 1.

We may picture the construction of strings as shots on a golf course (see Figure 24). It is important to note that 11 does not mean eleven; rather, it means 1 and 1. So the doubling operation according to the D rule results in doubling the number of symbols—in effect, a primitive

---

* 'Shut' is an example of a verb that does not alter at all in the past tense.

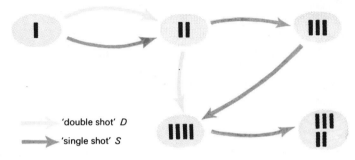

'double shot' *D*

'single shot' *S*

Figure 24

number system. The 'research' problem Rosenbloom has set twelve-year-olds is this: to determine par—the fewest number of 'shots' (that is, applications of one or other rule)—to arrive at a certain 'target' string. Let us work out par to reach the fifth 'hole' (labelled 11111). Using *S* shots only, we can hole in four shots. Our string is *SSSS*, in fact. But by making use of a *D* club as well, we can shorten the string to three shots: *SDS*, which is identical in outcome to the string *DDS*. On the first shot, *S* and *D* are interchangeable. Although, in general, *SD* is not the same as *DS*; or, as mathematicians would say, the *S* and *D* shots are not commutative. Rosenbloom has also posed this two-iron golf game to ten-year-olds as a mathematical research problem, which the reader might like to try: derive a general rule for finding par for a string of any length. For instance, par for forty-one, written as a string of forty-one 1s, is seven shots: *DDSDDDS* (or, equally, *SDSDDDS*). The secret is to write the shot number in binary: 41 becomes 101001 in binary. Then, ignoring the initial 1, for the first 'tee', we count the number of remaining digits in the string (here $6 - 1 = 5$) and count the number of additional 1s in the string (here $3 - 1 = 2$); we add these resulting numbers ($5 + 2 = 7$) to give par.

The counterpart of this golf game in linguistics is the transformational grammar developed by Noam Chomsky at the Massachusetts Institute of Technology. Suppose *A* is the present tense, first person singular of an intransitive verb 'to *A*', such as 'to run' or 'to sleep'; then 'I *A*' is a sentence. There are golf-like rules we can make up to transform this sentence into 'Are you *A*ing?' Recently George A. Miller of Harvard's Cognitive Centre has created a similar linguistic game-situation using strings of symbols and rules for combining them in his *Project Grammarama*.

The original word game

This game—Roger Brown's 'original word game'—resolves itself into naming things and operations; it consists of sifting relevant attributes. To a child, even a name is an attribute. We might well regard the grille, body-line and sound of its klaxon as attributes of a Ford 'Thunderbird', but not the name 'Thunderbird' itself. Yet to a child the name is indissolubly linked with the thing named. Piaget has made great play of this finding though it was known much earlier to the Russian, L. S. Vigotsky, who in 1939 wrote:

'When children are asked whether it is possible to replace the name of one subject with that of another, for instance to call a cow ink, and ink a cow, they answer that it is entirely impossible, because ink is used for writing and the cow gives milk. The exchange of the name means for them also the exchange of the qualities of the objects, so close and inseparable is the connection between the two.'

Learning a first language is a problem of attribute analysis—only the learner does not know what the attributes are; he doesn't know what is significant. The Navaho Indians, as Benjamin Whorf found, can and do discriminate between the 'a' in 'mat' and the 'a' in 'mad', or between the 'p' in 'nip' and the 'p' in 'pin'. A European would never hear such sounds as significant, and would not normally distinguish them— except unconsciously, as for instance, in placing a foreign accent. Man's ability to so discriminate is due in part to the extraordinary physical nature of the ear. As an instrument it is almost 'dead-beat'; it does not resonate like a piano.

Recent research by Chomsky and others suggests that children are born equipped with an innate 'knowledge' of language.* 'It looks as though at some level', John Davy wrote in the *Observer*, 'all languages share a common deep structure, which, if formally defined, would represent a "universal grammar". The difference between languages would then be expressed as distinct sets of rules for transforming the common deep structure into a variety of surface structures'— that is, different languages. The mathematical golf game gave us a glimpse of such transformations.

The very fact that children can learn English, Hungarian, Finnish, Hopi or Chinese, all with equal facility and sometimes under the most

---

* Chomsky's postulate of a 'deep structure' raises a problem as yet unresolved: the human brain has only developed language over some ten thousand years, which seems hardly long enough for a 'deep structure' to have evolved.

gigantic handicaps—deaf-and-dumb parents, impoverished back-grounds, partial sightedness or deafness—points to one thing only: that there seems to be a cultural heritage of language, *tout court*. Another significant fact emerges from research into language learning: as Davy says, 'childish' language is by no means an imperfect copy of adult language, 'but is a "language" in its own right'.

The salient outcome of these researches was the discovery that children between a year and a half and two years old go through a distinct phase of uttering two-word sentences. The words they use come from two classes; they are not selected willy-nilly, it seems. The first class, the smaller, consists for an American child of words which are verbs, pronouns, adjectives and exclamations. A typical set of such words might be 'all-gone, bye-bye, my, see, want, pretty, big, good, bad, little, hi'. The other class is formed of nouns and pronouns—girl, boy, me, shoe, Mummy, Daddy, milk, cup, and car. When the child makes sentences he picks a word from the first set and makes a string with a word from the second set *in that order*. Thus he might say 'all-gone milk' but never 'milk all-gone'—neither sentences likely to be taught by the parent or to be overheard being used by adults.

A mathematician would say that the child generates strings of words, to make sentences, as ordered sets—in fact, as ordered pairs of phrases. A linguist sees this rule structure for forming sentences as a strategy. The child's strategy of verb-like word followed by a noun is not the only strategy man uses in speech. The difference between a noun and a verb to a Hopi Indian, Whorf discovered, is that a noun is more permanent than a verb. We might be tempted to suggest that a non-linguistic situation (such as a buffalo charging) which calls forth the utterance of a verb is something that is perceived; a noun (such as 'tomorrow') takes rather longer to grasp and may even require infer-ence and intelligent thought. But certainly nouns have an air of per-manence.

Verbs need not be structured according to time. The Navaho Indian uses a different 'tense' according to the degree of credibility he attaches to his statement. Thus he will frame the sentence 'Fred lit a fire!' if he believes it to be true, in one tense; if a remote possibility, in another; and if tantamount to barefaced calumny, in a third. The tense of the statement becomes, then, a measure of the credibility gap. Altogether, the tenses, which are indicated by slight inflections, comprise a strategy for speaking. The astonishing outcome, as Whorf saw it, is that *what* we think is largely dictated by the language which we use to think or say it in. But perhaps this should not surprise us. Certain languages have always had the reputation for conveying thought of one kind or other —French for analytical thoughts or romantic sentiments, German for hortatory commands and scientific laws, English for phlegmatic observations and pragmatic rules, and so forth.

Is mathematics a language?

We could draw a parallel in mathematics and suggest that certain kinds of symbolism and lines of thought lend themselves to certain types of mathematics. A simple card game for children described elsewhere* by one of the authors reveals the strategies of such structuring. Each card represents a simple sentence. It has three symbolic cues on it: an outer shape such as a triangle, circle or square, an inner shape of the same kind, and a background colour—blue, green or red perhaps. The outer shape indicates the subject of the sentence; for instance, it might be a triangle for Tim, or a circle for Carole. The inner shape represents the verb in the sentence: perhaps triangle for 'runs', and circle for 'eats'. The background colour indicates an adverbial phrase—red for 'by moonlight' or blue for 'very happily', and so on. For the child the game is to match a picture illustration, say, a scene in which 'Tim eats very happily' with its appropriate card. Instead of matching a non-linguistic cue with a linguistic symbol—as normally happens when a child speaks —he matches non-linguistic cues with non-linguistic symbols. In the nature of, and response to, these cues, resides a valid distinction between mathematics and language. We cannot touch or eat mathematical cues— they are, after all, insubstantial relationships formed in the mind; all we can do is picture them in symbols.

Language also differs radically from mathematics in one other respect: speech may seem to be presenting certain facts, but it may be expressing something very different. The mathematician who tells us he is not much good at maths does not really expect to be taken at his word! Far less the drunkard claiming he is a 'no-good'! His words are an expression of a sentiment which the astute listener will interpret as needing assurances to the contrary. Such a subtle, double-think mode of communication just has no place in mathematics. However, as a partiality for linguistic cues amounts to a kind of strategy for language learning, so a penchant for mathematical cues can also be a sort of strategy in learning mathematics. But what the equivalent of 'expres-

* Holt and Dienes, *Let's Play Maths* (see bibliography).

sion' in the language of mathematics is, is not at all clear (see section *The chop-logic of words* in Chapter 3).

It seems probable that the act of cognition in both mathematics and language reflects something of the Eureka-like quality of Helen Keller's sudden learning of the word 'water'. The scene of the young girl, deaf, dumb and blind from birth, as she let water from the pump flow over one of her hands, and as her teacher spelled out the letters on the open palm of her other hand, must stand as the supreme model of sudden cognition, indeed, of perception.

The problem of how we attribute our linguistic categories to forms has exercised the greatest minds. John Locke, for example, in 1660 recognized a triangle as forming a generic image of one, in the abstract. He saw a triangle as 'neither equilateral, equicrural, nor scalenous but all and none of these at once'. Bishop Berkeley, on the other hand, envisaged a mental image as having only those essential attributes which have been selected for attention. E. B. Titchener, the psychologist, in 1909 was not even sure if the triangle in his mind's eye had angles which joined or whether there were angles in it at all; such imagery seems to be reminiscent of the rat's triangle tests (described earlier in this chapter in the section *Space and shapes*). At a higher level still, in handling mathematical or scientific concepts it seems likely that we simply perceive things, instead of working them out attribute by attribute. G. Spencer Brown suggests a difference between recognition and analogy or inference: 'recognition' takes place when the cognition is 'snap', and the other two modes of cognition take a long time. According to him, clever people 'perceive' where the less gifted have to puzzle things out. When he looks at a cow through a hole in the barn door, he wonders how it is he knows it is a cow from the part he sees. By inference or by perception? Similar problems attach to the understanding of mathematics. And probably they will never be completely cracked.

CHAPTER THREE

The Unreasonableness of Logic

Scene: *A restaurant.*
Morecambe: *'There's the manageress.'*
Wise: *'How do you* know *it's the manageress?'*
Morecambe: *'Well, it's certainly not the manager!'*
MORECAMBE AND WISE SHOW

'What, comrade, is capitalism?'
'The exploitation of man by man.'
'And what is communism?'
'Why, the reverse.'

'A wealthy franklin sent his son to Oxford and, having him home for the vacation, asked him what he had learnt. "I can prove that two is three, father," replied the son.

' "There are two fowls in the dish in the hearth," said the father. "Can you prove that there are three?"

' "Indeed I can," returned the boy and, picking one up, said: "Here are two fowls; and one and two make three, ergo here are three fowls."

' "Very good," said the father. "Now give one of those fowls to your mother, and one to me, and you shall have the third for your supper." '
Sixteenth-century, anonymous
story from A Hundred Merry Tales (freely adapted)

Marxian logic

Groucho Marx, on giving up his membership of a Hollywood club, wrote to its president: 'Please accept my resignation. I don't care to belong to any club that will have me as a member.' Marx had thus given a new twist to a celebrated logical puzzle of Bertrand Russell's. The antinomy, as it is called, goes like this: 'In a certain town there is a barber who must shave all those people and only those people who do not shave themselves.' A harmless enough rule, we might suppose. But does the barber shave himself? Clearly, if he does, then he is breaking the rule; so he must not. And if he does not shave himself, then by the rule he must! Russell's paradox—a contradiction in rules—is much more than a maddening puzzle. When he posed it at the turn of the century, it revealed distressing cracks in the logical house of cards Victorian thinkers had so painstakingly constructed in keeping with their own architectural monuments. Because of Russell's antinomy, and others equally disturbing propounded by Jules Richard and Cesare Burali-Forti, mathematicians were compelled to revise drastically their ideas on logic. In *Through the Looking Glass* Lewis Carroll reflects the prevailing intellectual climate in logic before the logical thunderbolts of Russell and others descended on the unsuspecting contemporary logicians: 'If it was so, it might be; and if it were so, it would be; but as it isn't, it ain't. That's logic.' What is the simplest way out of Russell's dilemma? It is to say that this barber cannot exist. It is even simpler to reiterate Russell's own advice to tyro scholars—not to study logic!

The main problem in grappling with the intricacies of logic is that most examples are couched in disarmingly everyday terms. Alfred Tarski discussed the logical snares that may entrap the unwary user of everyday statements to buttress his logical deduction. The reader, then, may be forgiven if he believes that logical thinking is but a deepening of his own intuitive day-to-day thinking. This is palpably untrue. Logic has precious little to do with everyday language as we shall see in the section *The chop-logic of words*. The nearest 'real life' comes to being logical is in the courtroom, perhaps one reason for Lewis Carroll's fondness for conundrums with a legal flavour.

In a play of the late 1960s, when in a court scene the judge asked the prisoner in the dock how he pleaded the accused replied, all too humanly, 'I am not sure,' whereupon the judge tartly reminded him that in law one is either guilty or not guilty. If not two-faced, the law is emphatically two-valued. It is like flipping a coin which invariably lands heads or tails; we discount the chance of its landing on its edge; should we recognize this remote possibility, we must assume a three-valued logic. It is also like switching a switch. In two-valued logic—which we shall assume for the rest of this chapter when we refer to logic *tout court*—there is no allowance for the switch getting stuck in a halfway position. Logic, then, is not about truth or falsehood, as we generally understand the words; it is about consistency—whether things relate without contradiction. Another unfortunate confusion over logic arises in the frequent use by logicians of the words true' and 'false'. In everyday terms logical propositions may be tantamount to arrant nonsense. (One thing should be made clear here: not every statement we make is a proposition in logic. To be so it must be capable of being true or false. For instance, an expression or command does not qualify.)

Tweedledee logic

That a proposition cannot be partly true or partly false implies that logic operates by virtue of the excluded middle: *a*, whatever it may be, cannot be not-*a*. Here, at least, logic is a respecter of reality. The dichotomy of *a* and not-*a* leads to the idea of logical implication. It must be emphasized that an implication is not in itself a piece of reasoning; it is merely a property. An example of a logical implication is 'If it snows, I shall wear goloshes'—or, more formally, 'If *a*, then *b*.' What is the connection between this and the excluded middle? Simply this, in specific terms: 'Either it snows or it does not snow' is a comprehensive insurance policy that covers all eventualities. 'Either it snows, in which case I wear goloshes, or it doesn't snow'—when I may or may not wear them; the rule is not specific about this aspect of my sartorial behaviour. So we cover all eventualities by saying: 'Either I wear goloshes' (implying that it is snowing) 'or it isn't snowing.'

No trivial exercise in hair-splitting, this; few people realize the full implication of the condition, 'If it snows, then I wear goloshes': that if I *do* wear goloshes, it does not actually have to be snowing! On this point logic is adamant. 'If we commit a crime, we may be punished.' But, in totalitarian countries at least, it is not true to say 'If one is punished, one has committed a crime.'

John Venn's diagram

In the 1870s an Oxonian logician John Venn devised a special diagram to illustrate logical propositions. The now-famous Venn diagram in Figure 25 shows our universe of propositions *a* = 'it snows' and

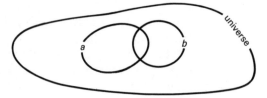

Figure 25

b = 'I wear goloshes'. The outer frame defines the set of things we are talking about, the so-called universe of discourse. Outside the frame there is no information to be sought or found. Between the double beer rings and the frame there is a no-man's land occupied by everything that is neither a nor b nor both. The space represents the combination of the propositions 'it neither snows nor do I wear goloshes'. We have drawn potato outlines for the Venn diagram but their actual shape is quite immaterial. The notions of logical inclusion and exclusion require for their representation only the concepts of 'inside' and 'outside' as used in topology for their depiction.

The first statement 'a or not-a' covers the entire universe of discourse. So we look for a valid conclusion anywhere within the Venn diagram. The second statement 'If a, then b' needs more thought. The only part of the diagram that corresponds to the statement, when a and b hold (when 'it snows' and 'I wear goloshes'), is the lune where a overlaps b. We shade the lune to show this to be a valid area in which to look for a conclusion. But what if a does not hold? Now not-a implies that b either may or may not be true. So we shade everywhere outside the a ring (see Figure 26). We use our tool next to represent by shading

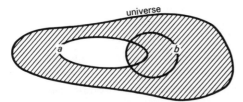

Figure 26

where the conclusion 'Either b or not-a' holds (see Figure 27) and see if this coincides with the shading for 'If a, then b'. If it does, then we have established the link. First, we shade everywhere that refers to b—that is, the enclosure indicated by the b ring; then we shade everywhere that refers to not-a—that is, everywhere outside the a ring. And, expectedly, we find we have shaded the same area as before.

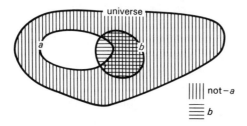

|||| not-a

≡ b

Figure 27

Torts and syllogisms

The Venn diagram may have few uses, but two examples, an everyday legalistic and a more macabre, theological one, will show its value in demonstrating logical principles. 'Although a tort is a civil injury', *Salmond on Torts* declares, 'not all civil injuries are torts.' The point is, if a person may not recover damages for loss sustained, then a civil injury cannot be classed as a tort. On a Venn diagram this legal assertion looks like this:

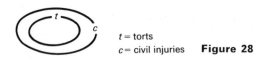

t = torts
c = civil injuries　　**Figure 28**

In Figure 28 t stands for torts and c for civil injuries. The diagram is not perfectly general; it only suits this solution where one ring is totally enclosed inside the other. In our words, this means that all torts are civil injuries; in legal words, 'a tort is a species of civil injury or wrong'.

There is a linguistic difficulty involved in the idea of 'either/or-ness'. Suppose we select cards from a pack according to the simple rule that we remove either Court cards or Hearts. We will end up with all the Court cards *and* all the Hearts, including the Court cards in Hearts. How to resolve this verbal paradox? We start out with 'either . . . or . . .' and end up with 'and'. The answer is simple but hard to keep in mind: the rule for forming the set—the selection rule of choosing either Court or Hearts cards—refers to the attribute of any single card in that set. Any card we pick in that chosen set must be either Court or Hearts. However, the entire set, as opposed to any card-member of it, will include Court cards *and* Hearts. So the words 'either . . . or . . .' refer to individual cards; the word 'and' to the entire set. There is another linguistic snag. Mathematicians use the word 'or' inclusively; popularly we use it exclusively. If we have a choice of going to the cinema or to the races, it is not usual—let alone feasible—to go to both!

Logic has not always had such an open face. It has had its darker side, as with the Elizabethan's sinister application of the syllogism to ferreting out hapless, if not witless heretics. Let us look at the case of John Gerard, the Elizabethan Jesuit missionary. He was questioned by the Dean of Westminster, and the following account comes from Philip Caraman's excellent translation from the Elizabethan Latin:

'The old Dean was irritated. "I will prove that you are a heretic," he said.

' "You won't be able to prove that," I retorted.

' "I can prove it," he said. "Whoever denies the Sacred Scriptures is a heretic. You deny this [the Sacred Bible done into English by the Protestant reformers] is the Sacred Scriptures. Ergo."

' "That's no syllogism," I answered. "It descends from the general to the particular and it contains four terms."

' "I could make syllogisms before you were born," the old man answered.

' "I am prepared to admit that," I said, "but the one you have just made isn't a syllogism at all."

'But other people broke in; they had no intention of beginning a disputation. They just threw questions at me in the hope that I would say something I did not want to say. Eventually, they sent me back to prison.'

Shorn of its menace, the syllogism can be presented as a Venn diagram (see Figure 29). From this it is as clear as daylight that the old Dean

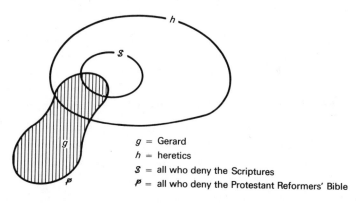

g = Gerard
h = heretics
S = all who deny the Scriptures
P = all who deny the Protestant Reformers' Bible

Figure 29 John Gerard—heretic or not?

had not proved his case against Gerard. For g (for Gerard) can be found anywhere inside the shaded ring representing 'all who deny the Protestant reformers' Bible'—which is not in all respects the same as the Catholic Bible, the Sacred Scriptures. The conclusion to be drawn is that Gerard is not necessarily a heretic—that is, he doesn't have to belong in the h ring. Gerard was lucky and influential enough to escape from his prison, The Clink, and lived to see another day. Perhaps the only logical conclusion to be drawn is Voltaire's. In his study of Louis XIV he asked how it is that while the differences among pagan religions could be peacefully settled, Christian disputes have inevitably led to bloodshed and war.

Lewis Carroll's diagram—updated

The Venn diagram suffers however from inflexibility. At most it can be used with a mere four variables with any comfort. But an elegantly flexible variant exists in the development of Lewis Carroll's sorting picture. Nowadays this kind of diagram is called a Karnaugh or Veitch–Karnaugh diagram and is used in industrial planning (see Figure 30).

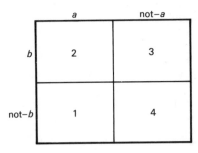

Figure 30

As before, there are four regions: (1) a but not-b; (2) both a and b; (3) b but not-a; and (4) not-a and not-b. And we can increase the number of variables at will; for example, the diagram below shows three variables a, b and c (see Figure 31) from which it is clear how the diagram may be extended to accommodate any number of variables.

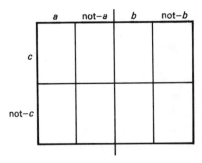

Figure 31

Figure 32 shows the Carrollian version of the 'snow-goloshes' Venn diagram (see Figure 27).

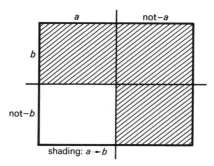

shading: $a \rightarrow b$

Figure 32

We trace the argument, more briefly this time. By statement (1) '*a* or not-*a*' is true, so we look all over the diagram. Statement (2) 'If *a*, then *b*' means we must shade the top left corner. And if not-*a*, what then? As before, the truth may embrace *b* or not-*b*; so the right half of the diagram needs shading. To see that statement (3) 'Either *b* or not-*a*' gives identical shading, we first shade the top half of the diagram for *b* (all of it) and then the right half for not-*a*.

In his diverting books on recreational mathematics, Martin Gardner has long advocated the use of such diagrams for solving 'Sunday paper' logical puzzles. But as Veitch–Karnaugh diagrams they also have a well-merited and respectable application in mathematics.

Polish notation

This notation—so named because it was introduced by the Polish logician Łukaciewicz—is a symbol system that avoids the use of parentheses. It does not resemble a conventional algebraic notation. In spite of this fact or perhaps because of it, it appeals to children, and it is a feature of the popular game, WFF-'n-PROOF, devised by Layman E. Allen in the US. The notation is also ideally suited to the needs of a computer. Łukaciewicz, logicians recall, was the first to show that the logic of whole sentences (like our 'I wear goloshes.') is a legacy of the Stoics and *not*, as believed, of Aristotle: he dealt only in sensible terms (such as our torts and civil injunctions or Gerard's indictment for heresy). Aristotle's logic is powerless to handle either sentences or nonsense terms such as 'green gravity' or 'square circles'.

The notation is simple: it involves the letters A K C N for the four logical connectives: A for the disjunction, K for conjunction, C for implication, and N for negation. A*ab* means either *a* or *b*; the A we can think of as standing for 'alternative'. K*ab* means both *a* and *b*—the K standing for 'konjunction' perhaps. C*ab* means if *a*, then *b*—the C standing for 'condition'. And finally, N*a* means not-*a*—the N meaning negation. The connectors A, K, C, and N must be written directly in front of the symbols for the variables (attributes, say) to which they refer. It may be that one of these variables is a composite one, formed by the prior connection of two other variables. Thus, let us introduce a third variable *c* (for 'I go out'). Then we could say: 'If it snows and I go out, then I wear my goloshes'—that is, in symbols, 'If *a* and *c*, then *b*.' In the Polish notation, we write this:

C K*ac b*.

To translate, symbol by symbol, the K (for both . . . and . . .) extends two symbols along; it embraces the *a* and the *c* only. Now we are talking in terms of playing with symbols; they are acting as mini-computers for us, thus saving us brain-fag. So K*ac* now behaves like a single symbol, which is linked to *b* by the pre-symbol C for implication or condition.

In our Venn and Carroll diagrams the various regions may be labelled in the Polish notation (see Figure 33). Readers for whom the idea of an

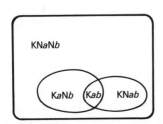
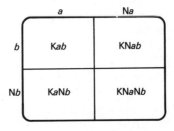

Figure 33

infallible machine has appeal may enjoy seeing the way implications are combined. Take, for example, the implications 'If *a*, then *b*'—that is, in Polish notation C*ab*—and the implication 'If *b*, then *a*' or C*ba*. The implications can be shown on a Carroll diagram (see Figure 34). To combine the implications, we must mentally superimpose the one diagram on the other, imagining they are drawn on transparent sheets. We then retain in our mind's eye only those parts that are double-shaded—that obey both implications. A chequerboard pattern of

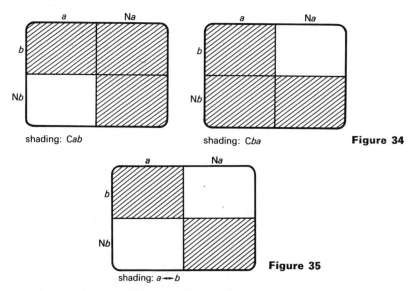

shading: *Cab*

shading: *Cba*

Figure 34

shading: *a→b*

Figure 35

shading results (see Figure 35). This is the shading typical of the double implication 'If and only if *a*, then *b*'; there is no room for doubt about what happens if it is not-*a*. If we were to say 'If *a*, then *b*' and 'If not-*a*, then *b*', we would find the complete Carroll diagram shaded; in this case we would have established a valid piece of reasoning true for the whole universe of discourse—what is known technically as a *tautology*. If the reasoning did not hold anywhere in the universe (no shading anywhere in the Carroll diagram), we would have an *antilogy*. Russell tired of mathematics on the grounds that he considered it merely the study of tautologies.

Finally, we can derive a fundamental and intriguing pattern in negating propositions. A glance at the Carroll diagram for the implication *Cab* reminds us of the shading for, what comes to the same thing, the statement 'Either *b* or not-*a*' or *AbNa*. The unshaded, lower left corner suggests a crisp definition. Why not call the shaded part what is *not* the unshaded part? The unshaded part represents 'both not-*b* and *a*' or *KNba*; so what is not unshaded must be *NKNba*, which the eye tells us is indistinguishable from the shaded part. Negating has the effect shown by the arrows below:

N(KN*b a*)

A *b* N*a*

To emphasize the mini-computer aspect of the symbols, we have written the formulae as a machine would write them. The N outside the top formula can be regarded as producing the following effects on the symbols after it. N changes K to A. N wipes out N, and this seems reasonable when we think of the frequency with which teachers correct children's misuse of the double negative. And where no N existed inside the bracket, the N outside the bracket puts an N in. Such are the simple computing rules. And they invariably work. Though not devised in this way, they are well known as 'De Morgan's rules'.

Logical relations

The exchange between the comedians Morecambe and Wise, quoted at the beginning of this chapter, relies for its humour, believe it or not, on a basic logical necessity: the need to define what we are talking about or, technically, the *universe of discourse*. The humour depends on the absurdity of confining this universe in a restaurant to two people only, the manager and the manageress. What makes us laugh is the *bisociative act*, to use Koestler's phrase, of having logic bestraddle two worlds, the one full of everybody we know, the other of two people only. Granted, then, that we have settled on our universe of discourse, we must attend to the properties of the things in it and their relationship to one another.

In all logic and much modern mathematics runs a familiar strain: the three properties of relationships. Perhaps the most plainly evident property is that of *symmetry*. To take a simple example, Ann may be the sister of Sue; so Sue is Ann's sister. The 'sister' relationship (in both ordinary and technical senses of the word) is, then, symmetrical. If we call that relationship by the letter R, we may write $a \, R \, s$ and $s \, R \, a$, where a stands for Ann, s for Sue, and R for the relation 'is the sister of'. Of course, the relationship might not extend both ways: Ann might be the sister of Bob (b). We can still write $a \, R \, b$, but we cannot write $b \, R \, a$ because b (Bob) is not a's (Ann's) sister. Instead we may write $b \, \bar{R} \, a$, where $\bar{R}$ means 'is *not* the sister of'. This relationship is not symmetrical, or, at any rate, not for this particular set of people, Ann, Bob, and Sue. For a different set (all sisters) there would be, of course, a symmetrical relationship.

Secondly, relations may be *reflexive*. For instance, in the class of synonyms, each word 'has the same meaning as' any other word in that class. So we could write 'small' R 'little', where R stands for the words in quotes in the last sentence. We could also write 'small' R 'small', for the word 'small' has the same meaning as itself.

Thirdly, relations may be *transitive*. A good example is the relation 'is harder than'. Diamond is harder than glass, which is harder than butter. So we write diamond R glass, and glass R butter, where R now means this new relation 'is harder than'. From this we may deduce the verifiable fact that diamond R butter—diamond will scratch butter. Some relations are not transitive, for example, the liking relation 'is a friend of'. Ann likes Bob, who likes Sue; but this is no guarantee that Ann is a friend of Sue—the liking relation does not carry over; it is not transitive.

All three properties of relations depend on the set of things or people chosen, and this is why a mathematician must specify a relation *with respect to* a given set. A mathematician's interest centres extensively around relations with all three properties. These are known as equivalence relations, the most obvious of which is 'equals' (having the same number as). A less obvious relation is one of likeness, for instance, the sartorial relation 'wearing the same colour as'. Our first impression is to say this is an equivalence relation . . . until we consider an example. Ann (who has nothing purple on) wears a red beret that matches Bob's tie; and Bob's purple shirt is the same colour as Sue's tights (although she has no red clothes on). So the relation links Ann and Bob, and Bob and Sue, but does not carry over from Ann to Sue. Thus, this likeness relation is not transitive. And so it cannot be an equivalence relation. If, however, Ann, Bob, and Sue only wore clothes of one colour, then this likeness relation would be an equivalence. For if Ann wore the same colour as Bob, who wore the same colour as Sue, then Ann must have worn the same colour as Sue—transitive, reflexive, and symmetric, and so an equivalence relation.

Relations can be sought between things and things, between things and sets of things, and finally between sets and sets. To establish a set of things we look for a common property. Thus all red clothes could belong to the set of red things; each member of the set has the property of redness. The study of such properties (mathematical rather than physical, as a rule) leads on to logic; while the study of sets leads to arithmetic and algebra. The reader can see, then, why the study of sets is on most school modern maths curricula, and why—because of the difficulties of teaching it—logic is not!

Clearly, the statement 'Either not-*a* or *b*' has the same logical force as the implication 'If *a*, then *b*.' (The implication is, however, merely a property, not a piece of reasoning in itself.) But not so clearly, perhaps, the relation between the two statements turns out to be another example of an equivalence relation. They have the same logical worth and are thus *equivalent* in logic—as a pile of ten pennies and a 10p-coin are in currency.

Finally, a relation can be envisaged as a passive link established between two entities. The relationship is there: it does nothing to alter the *status quo*. An operation, on the other hand, is something done to two such entities to make a third entity. To draw a very crude analogy, a relation is not unlike a family relationship: it's simply there. But, an operation is more like a marriage: two parties join in matrimony to produce (perhaps) a third party.

Psychology and logic

Piaget pursued somewhat inconclusively the application of this notion of relations to the comparison of logic with psychology. The psychologist's job, of course, is an empirical one; it is to suggest likely hypotheses (out of a host of possibles) and test them. Unlike physics, logic is not an empirical science; it starts from self-evident axioms and provides rules for detecting the validity of arguments and the truth of conclusions. In describing the study of relating mental structures to logic, Piaget coined the word 'psychologic'. Possibly, he did so with the aim of establishing in psychology an equivalent to mathematical-physics in physics. He conceded that much of human thought is unformalizable in logical terms, as we shall see later in this chapter in the section *The chop-logic of words*. Really, only mathematics is totally formalizable. Indeed, the American psychologist, F. H. Allport, went so far as to suggest that mental structures defy mathematical and logical treatment.

Perhaps the most puzzling feature of logic is its dependence on the disjunctive concept. We have already noted primitive man's inability to cope with it; for example, the Papuans, once head-hunters, are nonplussed by the prospect of alternatives. Guilford and Bruner have both stressed the antipathy man evinces towards the concept. As an instance of this, let us think of a gathering—a set—of distinguished men from all walks of life, such as may be seen at the Savage Club in London. Here is our universe of discourse (the accent on the last word, perhaps). Within it there is an *élite* of all those Savages, as they alarmingly call themselves, who are either mathematicians or race-goers. The situation may be shown on a Venn diagram (see Figure 36). The *élite* is literally an in-group—or as a mathematician would say a subset. It can be sought *anywhere* inside the two shaded rings. We could set ourselves this sort of problem; to find out if a Savage belongs to the *élite* group or not without actually asking him outright. We may ask two kinds of

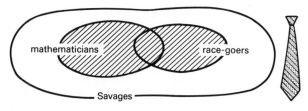

Figure 36

question only: what a man does or what a man likes. Evidently, anyone who is a mathematician *and* a race-goer is in. But if we discover that a man is not a mathematician, this does not mean he is not eligible; nor would discovering someone who does not approve of horse-racing. Both might be positive instances of membership of the *élite*.

Human nature usually ignores the negative instances in favour of positive ones. This blind preference extends even into science. A contemporary example is provided by the clinical psychologist who may follow this calling by virtue of being a qualified psychologist or of having worked, without paper qualifications of any kind, in a mental hospital or similar institution. The overall qualifications, then, are disjunctive. To regularize the situation, which evidently gravely concerns clinical psychologists, they have established a qualifying exam, which in effect renders conjunctive the entrance qualifications.

Arguably, the decision to couch the Ten Commandments in terms of what we may not do, might be cited as a case in favour of a predilection for the negative—until, on second thoughts, we realize that it is the sin or crime we concentrate on, and not the negative instances (different from the *negation*, of course) of the commandments! It is the job of the law to decide about the negative instances—wherever the transgression is not clear cut. A case in point is the Seventh Commandment: in law it is possible to find against one party of an adultery suit but not the other partner! A non-symmetrical relationship, indeed.

Another, older example, is G. K. Chesterton's Club of Queer Trades—peopled by all those creative misfits whom society cannot pigeon-hole into an existing job. Such men must create their own jobs because their talents and inclinations embrace quite disparate areas. If their abilities belong to disjoint sets (on a Venn diagram the rings do not overlap), then the possessor must pursue each activity in watertight compartments, like D. Jackson, the American physicist who also raced in the Grand National, or Will Hay, the comedian, who was also a skilful amateur astronomer, or the polymath Leonardo da Vinci. The supreme master of the disjunctive concept is the successful inventor who, like Autolycus, takes unconsidered trifles from the kitchens of widely different sciences and arts and creates a confection never seen before. The pseudo-medical science of homoeopathy, where a very little dose of what does the harm is administered to cure a person on the 'hair of the dog' principle, highlights the maddening element in the disjunctive concept as far as scientists are concerned. A controlled dose of snake venom, for example, can cure someone bitten by a poisonous snake. The conclusion, then, in the words of the quack, is that snake venom 'either kills or it cures'. But there is another side to this coin, which is to look at disparate scientific results for the common feature —the shaded, overlap region on a Venn diagram. The scientist finds, for example, that insulin subcoma, electroshock therapy and narcosynthesis all produce an improved state in schizophrenics. As a result, he is immediately tempted to look for the common factor in the hope of finding a conjunctive ('and') method of treating schizophrenics; the classical cause-and-effect notion fits this category.

In *A Study of Thinking* Bruner suggests that when confronted by disjunctive problems, a solver usually adopts one of two incorrect strategies. Using our example of the Savage Club's *élite* group, let us suppose that a questioner is intent on determining the rules of membership. Answers to his questions reveal that of the members of the *élite* he asks, some are mathematicians, some race-goers and, puzzlingly, some combine mathematics or race-going with interests in the arts and so forth. What does our questioner do then? If he falls for one false, yet popular, fallacy, he will decide that all these men qualify for whatever reason he has heard the majority of times. It might be that he has questioned mostly playwright-mathematicians (like Georges Feydeau, the writer of farces), and only a few race-going mathematicians. By the 'majority' fallacy, he concludes, incorrectly, that the entrance rule is 'to be a mathematician and/or a playwright'. Or he may adopt a better strategy: he may get as far as the correct qualities, yet *erringly* plump for the rule 'race-goer *and* mathematician'—in deference to our penchant for the conjunctive. This penchant reflects the fact that in normal speech we use 'or' in the exclusive, rather than the inclusive sense, whereas in mathematics the inclusive sense is preferred.

Small wonder, then, as Piaget found, that children come to logic late in their school life. But when they do, there are certain aspects they should be able to grapple with—which we will discover in the next section in novel guise.

Scientific theory and logic

Because the immediate impression that logic gives is one of extreme rigour, it is deemed a subject difficult to learn. Yet the difficulties are more often covertly semantic than patently logical. This is shown most clearly by scientific theory which is shored up by logical syntax and semantics. Logical syntax concerns concepts of formal truth and falsity and the regulations that govern the activity of deduction. Semantics of more recent origin,* now more fashionable, deal with the theory of meaning and the notions of material truth and falsity. Deduction itself is simply the unravelling of a tautology. It is a way of saying the same thing in two different ways. Peter Medawar quotes H. A. Rowlands who put the matter admirably:

'Deduction obeys a Law of the Conservation of Law. It is only man's frail mind that needs Euclid to work out all his theorems, which should be perfectly evident, indeed deducible at a glance from the formal axioms. A Mechanical Brain could do the job as well.'

In one sense the rigours of deductive proof are curiously overrated. Logic proves to be quite a lenient discipline. For when it is said that one statement (*p*) follows from another (*q*), two-valued logic admits any of the four alternative combinations—*p* true or false coupled with *q* true or false—except one: when the first statement *p* is true and the second *q* false. All that it guarantees is that the consequent of a true premiss should be true. The burden of disproof rests on this single combination. The dilemma of 'proof' is that in logic two 'wrongs' can make a 'right'. This is why scientific theories are not proved true but are merely accepted hypotheses until proved wrong. The burden of proof rests, as Karl Popper sees it, on the resistance of a hypothesis to 'disproof'. If what follows from a hypothesis is shown to be false, experimentally say, then the hypothesis is no good, and the deduction is false in logic.

The asymmetry of implication is as important in science as it is in semantics, and is central to the scientific method. Many experiments are designed as *null* methods—to confound a null hypothesis. Just as we use words imprecisely, scientists have degrees of belief in their accepted hypotheses. Unlike logical theorems, scientific hypotheses are tested by experiment. A scientist is (according to Medawar, referring to disjoint classes) 'not often lucky enough to deal with a crisp disjunction. An hypothesis is less often outright false than merely

* See the section *Symbol—servant or tyrant?* in Chapter 2.

inadequate, and not beyond the help of running repairs.' For instance, brown eyes in a population of people are the phylogenetic outcome of the existence in their gene cells not only of dominant brown alleles but also of recessive blue alleles. This is not a case of a clear-cut disjunction with disjoint sets of characteristics, but of an overlapping conjunction of characteristics.

Science is the interplay of hypotheses and the logical expectations they promote, while hypotheses must be, as Kant showed, statements of such a kind that they *could* be true. Logic departs from semantics in that it deals with forms rather than meaning, including even the various allotropic form of nonsense. In everyday speech, the structures 'if . . . then . . .' and 'and' and 'or' can be used interchangeably. The threat 'If you don't shoot them, I will shoot you' is in the classical logical form and reveals the semantic difficulty already referred to. It is the understood thing that if the antecedent is complied with, then 'I' won't shoot 'you'. However, this does not logically follow, as American troops in Vietnam were quick to appreciate when their officers made the very same threat: they simply took no chances and shot the officers first! A case of questionable discipline but impeccable logic.

Syntax, too, fogs the semantic issue. Take the direct implication 'If you move, I shoot.' It can be rephrased to offer two alternatives 'Don't move *or* I shoot,' which is not commonly recognized as an implication. The construction establishes, albeit colloquially, the logical equivalence of an implication and an alternative or disjunction. Again the implication can be rephrased using the conjunction 'and' without the negative, as in 'Move and I shoot.'

Combined with the widely held belief in the stringency of logical implication, which as we have seen is not so rigorous as all that, certain sentence constructions conspire to make logic difficult to handle. Indeed, many everyday examples such as can be culled from textbooks may be more of a hindrance than a help to understanding.

Magnetic logic machine

It is possible to enlist the aid of two physical phenomena—magnetism and gravity—to create an analogue machine for the intellectual process of sorting in establishing a logical implication together with its equivalent disjunction, the alternative, with the antecedent negated. The following magnetic sorter, which is based on the physicist's *mass spectrometer*, for sorting atoms of different masses, is a cross-sorting device. It sorts small and large ball bearings, both magnetic and nonmagnetic, say steel and plastic balls. It consists of a thin, sloping perspex plate under and near which is located a strong magnet. When ball bearings are rolled down the plate, the magnet deflects the magnetic bearings off a straight-line track but not of course, the plastic bearings. These roll down a sloping ramp with small holes drilled in it—a second sorter—and the small plastic bearings, but not the large ones, fall through the holes into the reject bin.

Figure 37(a) is a composite plan and elevation of the machine. It is 'programmed' to sort bearings into an output bin according to the rule 'If small, then magnetic.' All others it consigns to the reject bin. Because the attribute 'small' occurs in the first part of the implication, a preponderance of small bearings might be expected to collect in the output

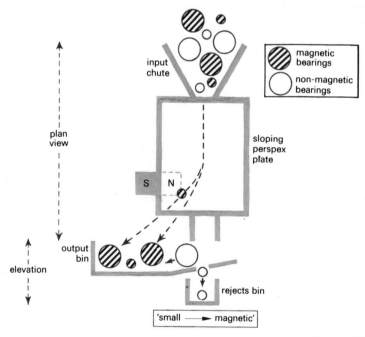

Figure 37(a)

bin. Not so, as experiment would show. The output bin will contain bearings that are either large (not-small) or magnetic, as we know from a knowledge of logic equivalence.

This is most clearly demonstrated by arranging the output bin as a

Carroll diagram and cross-sorting the contents according to the plan shown in Figure 37(b). The cross-sorting machine has the virtue of making plain in physically evident terms the distinction between the attributes of any single element (ball bearing) to be found in the output set (the bin) and the defining attributes (either large or magnetic) of the entire set, a point of some pedagogic value.

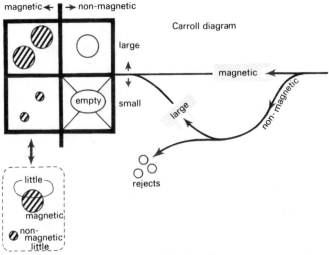

Venn diagram equivalent for output.(Shading indicates a non-empty set)

Figure 37(b)

It is interesting to note that sorting of this kind is the stock-in-trade of the zoologist classifying, for taxonomic purposes, such things as skeletal remains: he uses attributes of size, bone-structure, shape, and fittingness (like pieces in a jig-saw). Such sorting is made that much easier by the fact that phenotypes by and large correspond to genotypes. Where this correspondence falls down, as with the rabbit-like hyrax whose closest zoological relation is the elephant(!), taxonomy becomes that much more problematic. It is also worth mentioning that there are only two possible outputs—the selected set and the rejected set: this is because the situation is being treated as a two-valued logic. (This, too, is of some pedagogic importance.)

Paper logic machine

The reader may make very simply for himself a novel device which we call a 'paper logic machine'. It shows constructively how certain logical implications relate. Much as a constructivist mathematician would create ideas, the machine mimics exactly an utterly different kettle of mathematical fish, the so-called Klein group.

The reader may recall the four logical implications—the *identity* proposition, *converse*, *negation*, and *contrapositive*. 'Identity' is another word for 'restating the proposition' and reminds us of the Bellman in Lewis Carroll's *The Hunting of the Snark*, who based *his* proofs on threefold repetition, much as do demagogues, and cried: 'The proof is complete, if only I've stated it thrice.' The converse of the Bellman's cry would be 'If the proof is complete, I've stated it thrice.' By negating it, the Bellman's cry becomes 'The proof isn't complete, if I haven't stated it thrice.' (The word 'only' is omitted for the sake of elegance and clarity.) And here is the rub of chop-logic: many people mistakenly believe that a proposition *and* its negation follow. Finally, the contrapositive is 'If the proof isn't complete, I haven't stated it thrice.'

More briefly, we can demonstrate these logical connectives by calling the identity proposition 'If *a*, then *b*'. Then the connectives with our code letters are:

(*I*) *Identity*: If *a*, then *b*.
(*C*) *Converse*: If *b*, then *a*.
(*N*) *Negation*: If not-*a*, then not-*b*.
(*P*) *Contrapositive*: If not-*b*, then not-*a*.

Our definition of negation may strike the mathematically minded as unusual. For, conventionally, the negation of 'If *a*, then *b*' means 'Both *a* and not-*b*', whereas we mean here 'Either *a* or not-*b*'. That understood, the rest should fit together coherently.

Barbel Inhelder and Piaget have declared, in so many words that a child must be able to handle these connectives and the relations between them or he will never cope with secondary school maths or science. Be that as it may, the fact is we can combine two connectives in tandem to form a third. For instance, if we start with the identity proposition, negate it, and then take the converse of *that*, what do we get? Here are the stages and operations presented in three quite different forms of symbolism.

(1) *Verbal:*

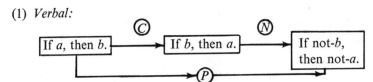

The reader can see that we may reach the last proposition in one jump through the contrapositive (*P*) alone.

(2) *Logical symbolic:*

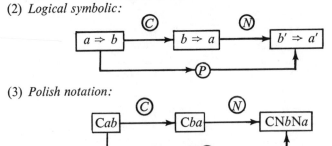

(3) *Polish notation:*

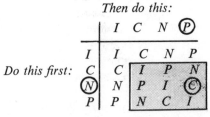

Now a rather peculiar fact emerges. If we start with a statement expressed in any of the four forms, and carry out first the *C* and then the *N* operation, we find that a *C* followed by an *N* is always tantamount to the single *P* operation. Even more briefly, we may write then:

$$CN = P$$

(If the reader feels more at ease with it, he can adopt a slightly different convention and write this equation as $C + N = P$; it comes to precisely the same thing.) By the time we have teamed up every possible pair of logical operations, we have sixteen results to hand. The neatest way to arrange them is as a multiplication table or an AA mileage table:

Then do this:

	I	*C*	*N*	*P*
I	*I*	*C*	*N*	*P*
C	*C*	*I*	*P*	*N*
N	*N*	*P*	*I*	*C*
P	*P*	*N*	*C*	*I*

Do this first: (rows *I*, *C*, *N*, *P*)

This table is of the Klein group, named after its inventor, Felix Klein, the celebrated German mathematician. The top row and first column are obvious when we think about them. In the top row, we read the first item as *I* then *I* (or *II*), giving *I*, which appears as the top left item within the table. The next item in the top row is obtained by combining *I* again, now with *C*, and since the *I* for the identity logical operation does 'nothing', in effect, we still have *C*. For the same reason the other two items in the top row read *N* and *P*. Similar reasoning gives the first column, identical with the leading column outside the body of the table. The really intriguing core of the table is the shaded square.

The items in the shaded square are come by in exactly the same way, but some interesting patterns emerge. The diagonal running from top left to bottom right consists of *I*s only. Doubling any operation, as the reader can check, has the effect of restoring the *status quo*—we are back to the identity proposition, as if there has been no change. Another feature is this: forgetting about the *I* item for the moment, if we pair any two of the other items, *C*, *N*, *P*, we get the third in the trio. Thus, *CN* gives *P* and *CP* gives *N*. As an example, we have ringed Ⓝfollowed by Ⓟ, which gives Ⓒ.

Now, as it happens, we can obtain exactly the same pattern of operations by doing the following things to a book. Pick it up and put it down as it was. We shall call this the *Identity* (*I*) operation; effectively it does not change the book's orientation. Spin the book round as if it were on a record-player turntable through half a turn. We shall call this the *Clock spin* (*C*). Then flip the book over onto its other side, as if opening a school desk. We shall call this the *New face flip* (*N*). Finally, turn the book as if turning its pages. We shall call this the *Page turn* (*P*). The reader has probably divined the reason for such strange names for these common or garden operations: the initial letters match nicely the letters standing for the logical operations. And quite remarkably, we find that, for instance, an *N* (New face flip) followed by a *P* (Page turn) has the same effect on the book as a *C* (Clock spin). Indeed, all the combinations of turns fit the pattern shown in our table. This similarity of pattern—one for one—is, the reader will recall, what a mathematician calls *isomorphism*.

It was this isomorphism that led the authors to experiment with a piece of card in the following manner. Cut out of a postcard any two obvious shapes. We chose a circle and a square (see Figure 38). These

front

back

Figure 38

represent two logical propositions—the Bellman's perhaps or one of the reader's own. So the circle stands for the proposition *a*, whatever it may be, and the square hole for *b*. We read the card left to right as 'If the shape on the left, then the shape on the right' or 'If *a*, then *b*'.

Now flip the card over by the *N* flip—we see the back of the card, shaded to show it is the 'not' side. If we do the *N* flip again, the card is in its original orientation again—a position at which, of course, we could arrive simply by doing nothing—that is, by doing the *I* operation. A glance at the table shows that this is so for the logical operations. Spin the card by a *C* spin and follow that by a *P* turn—Figure 39 shows

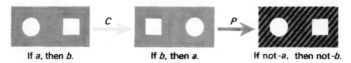

If *a*, then *b*.   If *b*, then *a*.   If not-*a*, then not-*b*.

Figure 39

the sequence of moves and positions. At this point we might look up *CP* on the table and find it is equal to *N*, or we might remember that any two of *C*, *P*, or *N* always gives the third operation. A suspicion might then form in our minds that somehow the *C* and the *P* turns could be replaced by *N* alone. And this turns out to be the case (see Figure 40)!

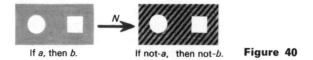

If *a*, then *b*.   If not-*a*, then not-*b*.   **Figure 40**

Further investigation confirms our astonishing mathematical discovery. All the sequences of logical connectives do indeed match, one for one, the runs of flips and turns with the postcard. The two systems have identical structures: they are isomorphic.

Figure 41 shows a bird's-eye view of the turns of the card:

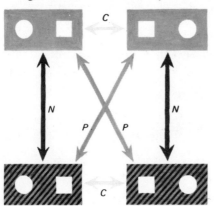

Figure 41 All the turns of the card

This is a map which shows us how to move from any orientation of the card (or for any logical connective, for that matter). We saw that starting with the card in 'neutral' so to speak, and using our shorthand, $CP = N$. What we did not note was that $CP = N$ wherever we start. This the diagram illustrates but the table does not. For instance, if we start at the bottom left corner and follow a C arrow (to represent the C turn), we get to the bottom right card; we then follow a P arrow and arrive at the top left card. But, starting once more from the bottom left card, we arrive at the same end-point along a single arrow only, the N arrow. Naturally the creative mathematician must be able to detect such isomorphisms—similar structures—in a real problem without the kind of symbolic help stage-managed here. In a real problem, too, there may be irrelevancies—what the information expert dubs 'noise'—which he has to learn to discriminate and ignore.

The chop-logic of words

Although regarding the logical connectives as a basic necessity for adult rational thought, Piaget recognizes that much adult thinking—other than mathematical thinking—is unformalizable in logical terms. Like John von Neumann, he concluded that the language of the brain is *not* the language of mathematics or logic. In normal thinking, generically speaking, propositions come first and axioms last—that is, the reverse order to logical processes. Furthermore, systems of information to describe thinking and intelligence, such as Guilford's cube-like model of our intellectual abilities, as he says, just do not 'lend themselves to a step-by-step treatment', which characterizes logic. Mental systems defy logical treatment, although one mathematician has attempted to pin them down by topology (see section *A model of the brain* in Chapter 5).

As with mental systems, so with language. Language is illogical. This may be demonstrated by logic in a somewhat Carrollian fashion by means of, aptly enough, a Carroll diagram. The following discussion is the outcome of a private conversation between Holt and Dienes, and stems from Dienes' current researches on language structures conducted with the linguist, J. M. Zemb. One of their aims is to investigate the effect of the logical connective of negation on a simple sentence and to detect if the sentence behaves as logic predicts.

Let us take the sentence: 'He came willingly.' The argument, or what grammarians used to call the subject, is 'He' and the predicate 'came willingly'. Keeping the argument constant, we negate the verb 'came' and the qualifying adverb 'willingly' in turn to see what happens. But, how *do* we negate the sentence? Do we say 'He did not come willingly' or 'He came not willingly'—more colloquially, 'He came un-willingly'? (The problem is akin to our earlier one of defining negation, near the beginning of the last section, *Paper logic machine*.) Neither sentence negates the original statement in the sense that 'He did not come' means the opposite of 'He came.' When put into a Carroll diagram, the gamut of possible sentences looks like this:

	verb	
	came	*not-came*
willingly	He came willingly	He did not come willingly
not-willingly	He came unwillingly	He did not come unwillingly

(left margin label: adverb)

Two striking facts emerge. First, not one of the variants means 'He did not come.' To a mathematician this suggests that the logical operation of negation does not distribute over the elements of the predicate in the sense that multiplying by numbers does; for example, $4(2 + 3) = 4 \times 2 + 4 \times 3$. (Here the multiplication by 4 distributes over the 2 and the 3 in turn.) Secondly, the lower right sentence has the same meaning as the one given at the outset (top left corner).

Filling in the Carroll diagram with the Polish notation produces this pattern:

	c *(came)*	Nc *(not-came)*
w *(willingly)*	Kcw	$KNcw$
Nw *(not-willingly)*	$KcNw$	$KNcNw$

What we are looking for is a sentence that means 'He neither came nor came willingly.' In symbolic notation this is $ANcNw$. From our earlier discussion of De Morgan's rules in the section *Polish notation*, we know that this is equivalent to $NKcw$, which decodes (perhaps ambiguously!) as 'He didn't come willingly.' The nearest we can get to it is the sentence at the bottom right: KN*c*N*w* meaning 'He both didn't come and un-willingly, too.' An uncolloquial sentence, to say the least. Now by De Morgan's rules this is equivalent to NA*cw*—'He didn't either come or willingly.' This *sounds* like what we want. The fact that it isn't should serve to put us on our guard against relying on the sense of words to sort out logic for us.

The same sort of dilemma presents its horns with interrogation as well as negation. Start with the sentence, 'He came', and negate it: 'He didn't come.' Then make a query of it: 'Did he come?' and also make a question of the negative: 'Didn't he come?' This is not at all the same thing as the negative question 'Did he not come?' The trouble is, that expression has entered the picture, and both logic and mathematics are powerless to cope with this. At least, however, the tools of logic make it much easier to analyse language structure—even if it is thereby shown to be grossly illogical or, at best, mere chop-logic.

CHAPTER FOUR

Models in Science

'Why, why is a banana curved?'
'If a banana was straight, then it would no longer be a banana.' HIT SONG OF 1920S

'Everything which can be the subject of scientific thought, becomes subject, as soon as it is ripe for a theory, to the axiomatic method, and thereby indirectly to mathematics.'
DAVID HILBERT

'[Man] tries to substitute this cosmos of his own for the world of experience, and thus to overcome it.'
D. N. CHOROFAS

'That there is more order in the world than appears at first sight is not discovered till the order is looked for!'
N. R. HANSON

'Advances in knowledge are not commonly made without some boldness and licence in guessing.' W. WHEWELL

When Newton saw an apple fall, he found
In that slight startle from his contemplation—
'Tis said (for I'll not answer above ground
For any sage's creed or calculation)—
A mode of proving that the earth turn'd round
In a most natural whirl, called 'gravitation';
And this is the sole mortal who could grapple,
Since Adam, with a fall, or with an apple.

Man fell with apples, and with apples rose,
If this be true; for we must deem the mode
In which Sir Isaac Newton could disclose
Through the then unpaved stars the turnpike road,
A thing to counterbalance human woes:
For ever since immortal man hath glow'd
With all kinds of mechanics, and full soon
Steam-engines will conduct him to the moon.

LORD BYRON: Don Juan
Canto the Tenth, I and II

Mathematician versus scientist

Many outstanding, creative scientists freely admit that they are not at home with mathematical ideas. Yet maths has come to be regarded as the 'Queen of the Sciences' or 'the language of science'. This apparent paradox might suggest that such scientists are like foreigners in the scientific world of their own making, and this clearly cannot be so. It is truer to say that there are large tracts of uncharted territory in any science that can be explored with quite unsophisticated conceptual tools; many scientists excel in the role of intrepid Colonel Fawcetts exploring the Amazons of the realms of science. None the less the most potent science can only be expressed in terms of maths, as we shall shortly see.

The fact remains, however, that a treacherous rift, badly camouflaged, keeps innumerable scientists and mathematicians at arm's length. In light of the subtly disingenuous attitude which each all too often has toward the other, the celebrated Snow–Leavis dispute seemingly fought by the more die-hard adherents of the 'sciences' and the 'arts' was by comparison a mere frog–mouse battle. Their points of view have been aired lately by that most able writer and scientist, Peter Medawar. But the problems that so often beset the scientist are like skeletons in a rarely opened cupboard, and indeed may well be related to the publicized turning away from the sciences by students.

The only frank reference to this unfortunate rift was in Nobel-prizeman James D. Watson's *The Double Helix*, which tells about his and Francis Crick's discovery of the DNA molecule of heredity. Before this work, Watson had been an ornithologist, or a mere 'bird-watcher', as his co-worker Crick often called him. This friendly joshing arose out of Watson's frankly admitted inability to handle Fourier series, so famous in mathematics. Crick felt himself in some way above Watson for his skill with mathematics. However, it is doubtful if a pure mathematician would regard such a handiness with advanced calculus as anything to boast about! Like the two winding staircases of a DNA molecule, affectionately dubbed 'Crick' and 'Watson', maths and science are intimately linked, for better and perhaps for worse!

In trying to portray the workings of natural phenomena, in this book, the authors are attempting an almost impossible feat—to describe mathematics without the use of too much mathematics! Should the expositor ever be questioned too closely, he has no option but to resort to mathematical symbolism to make his point. To the best of our knowledge, even major phenomena can only adequately be described in terms of mathematics. It does not matter what the question is—'How does the Earth orbit round the Sun?' or 'What are the origins of life?' All such questions need maths to answer them convincingly.

Far from swelling with pride, many pure mathematicians are positively upset by any demonstration of the universality of their subject. G. H. Hardy, for instance, the pure mathematician *par excellence*, could never reconcile himself to the prospect of his maths being of the slightest practical use. He once apologized, tongue in cheek, for a mis-spent life: 'I have never done anything "useful". No discovery of mine has made, or is likely to make, directly or indirectly, for good or ill, the least difference to the amenity of the world.' Nevertheless, in a letter concerning a problem on genetics, he tossed off a principle that has become known as the Hardy–Weinberg law, and that has, according to the scientist E. C. Titchmarsh, 'central importance in the study of Rh-blood groups and the treatment of haemolytic disease of the new-born'. This law, basic to genetics, scotched the 'common-sensical' view that when populations differing in some characteristic inbreed, succeeding generations show a 'blending' of genetic traits—colour of eyes, curliness of hair, blood group types, and so on. Thus if a population has, say, 60% blue-eye genes and 40% brown-eye genes, random interbreeding will not change these percentages. So much is known to geneticists. Hardy, who was English, and a German physician, Wilhelm R. Weinberg, independently transformed this into a mathematical model about genes.

A gene model

Eye-colour genes, to take an example, come in two forms, or alleles as the biologists say, B for brown eyes, the dominant gene, and b for the so-called blue-eyed, recessive gene. The B and b genes in the population of marriageable partners form a gigantic gene pool; we may think of it as a bottle of coloured beads, 60% blue and 40% brown. So 60% of the eye-colour genes in the population have the blue-eyed form—though not necessarily 60% of the population will have blue eyes as a result because the brown gene is dominant and, to make a pun, colours the issue. Suppose that p out of 100 germ cells carry gene B and that the rest, q out of 100, carry gene b. This means $p\%$ are the brown-eyed genes, and $q\%$ are the blue-eyed genes. Obviously,

$p + q = 100$, that is, 100%, representing the total population. Using fractions we could redefine p and q to give us $p + q = 1$.

Returning to our jar of coloured beads, we can represent the mating process by picking out pairs of beads. We consider the possible linkings (technically, zygotes) that may arise from the fertilization of B-bearing and b-bearing eggs by the same bearers in the sperms. Of the offspring in the population, BB (brown-eyed), Bb (still brown-eyed) and bb (blue-eyed) individuals occur at frequencies of p^2, $2pq$, and q^2, respectively. And that, in a particular nutshell, is the Hardy–Weinberg law.

In our example, $p = 40\% = 0.4$ and $q = 60\% = 0.6$. So $p^2 = 0.16$, $2pq = 0.48$, and $q^2 = 0.36$, and these numbers together add up to 1, again representing the entire population. The point of the Hardy–Weinberg law is that the frequencies of the B- and the b-bearing eggs and sperms is *still* p and q, just as for the parent generation. For the B-bearing eggs and sperms the frequency is

$$p^2 + \frac{2pq}{2} = p^2 + p(1 - p) = p^2 + p - p^2 = p$$

This is the same proportion as for the parent generation. The $2pq/2$ term arises from the fact that the Bb individuals will split between the B- and b-genes, and similarly, the b-bearing eggs and sperms.

Prime number machine

Hardy was never happier than when working with the sort of mathematics which Euclid established, surprisingly perhaps, about prime numbers. Euclid's proof that there are an infinite number of primes— numbers like 2, 3, 5, 7, and so on, that can only be divided by themselves and 1—was for Hardy a 'beautiful theorem', and part of 'pukka' mathematics, the mathematics of the working professional mathematician. The proof is set down in Hardy's elegant, cool prose in his autobiography, *A Mathematician's Apology*, which Graham Greene has described as 'moving, exciting, beautiful'. Less familiar is Euclid's Prime Number Manufacturing Machine, which appears in Book 9, Proposition 20 of *The Elements*, and might have come straight out of one of the avant-garde maths textbooks for school-children. The machine—Aladdin fashion—mints new prime numbers for old. An example will demonstrate its subtle mechanism.

Suppose we feed into the machine three prime numbers—say, 3, 5, and 11. The machine multiplies them together, to give 165. Then the machine adds 1, to give 166. Lastly it divides that result by the prime numbers in turn . . . until the result is divided into all its prime factors: here 2 and 83; the latter is a new prime. (Unless we feed in the prime 2, we always get a 2 in the output, because an odd number multiplied by an odd number always gives an odd number, and all the primes *except* 2 are odd.) We picture the machine as a contraption (see Figure 42).

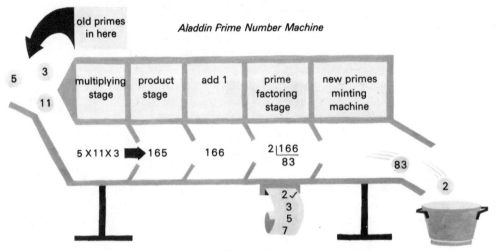

Figure 42

We might assume that the prime number theory would be safe from technological vandals. But not so. Even Hardy's beloved Riemann's Zeta function—a sort of ready reckoner formula for the number of prime numbers less than a given number—turned out to be falsely pure and found a useful, if unintentional, application in measuring the temperature of industrial furnaces. The point is this: not even Hardy's 'purer than pure' mathematics could escape the stigma of being useful by providing a 'model' of practical phenomena. Yet ironically, he enlivened one celebrated paper addressed to European mathematicians on 'A maximal theorem with function-theoretic applications', by taking cricket as his model. It can only be imagined what help his model afforded Hungarians or Russians grappling with his theorem! Despite its probable failure, the cricketing illustration was intended as just such a model as scientists use. The mathematical model is the most powerful tool a scientist has at his disposal.

Models—mathematical and otherwise

When we watch a fashion model or film star, we do not honestly believe they are models of 'us' any more than Shakespeare's plays are realistic portrayals of everyday life. The fashion model in a small way and Shakespeare in a big way both mirror life selectively and abstractly. Both are the glass of fashion, the mould of form; both offer handy shorthand symbols for handling the complexities of social fashion or human emotions. In much the same way, the scientist uses mathematical models; indeed, often he has no option.

The essence of a good mathematical model is that it should embody the 'bold ideas, unjustified anticipations, and speculations' which, according to Karl Popper, are 'our only means of interpreting nature'. The good scientist then puts his model to the test and exposes it 'to the hazard of refutation' or he does not take part in the scientific game—this is Popper's view again. Whether life appears to the scientist to be a bowl of cherries, or so many field equations, his aim is to use either model to gain control over his environment. His model is more likely to be a 'black box' than either of these two. Into this are fed values that have been metered on instruments and out of it pops an answer in terms of these values. More probably, his model might be purely mathematical in nature. All scientific laws, for example, are models (though not all models are laws!). Map-making, too, is a kind of model-making. Some models are perfect, they never let us down, and they throw a reliable intellectual bridge over the gulf between fact and supposition. Some are more flexible and, in consequence, more chancy. So-called analogue computers can be fashioned to be perfect (deterministic, in the jargon) or chancy (stochastic). Most computers are digital: they work in discrete steps while analogue computers work continuously and smoothly. Now this cliché distinction won't do, as Gregory has shown. The difference between them is basically an engineering, not a logical distinction. For example, cutting notches or 'stops' in a slide rule, the paradigm analogue machine will hardly turn it into a digital machine!

Lewis Carroll once made an amusing distinction between these two kinds of models. As the chanciest kind of model the Bellman's could not be outdone: it was 'A perfect and absolute blank!' In contrast, Carrollian cartographers in *Bruno and Silvie* experimented with larger and larger maps until they finally made one on a scale of a mile to a mile. The map, however, was never spread out. 'The farmers objected:

they said it would cover the country itself, and shut out the sunlight! So now we use the country itself, as its own map, and . . . it does nearly as well.' The maps John Speed and Christopher Saxton drew were essentially symbolic abstractions from reality, making little use of pictorial concepts. With today's more accurate Ordnance Survey maps, we can cast away these maps, as so many old clothes. (Their value none the less soars—not as maps but as precious historical documents.) And this is exactly what a scientist does when a particular model has outlived its usefulness.

As a supreme example of the scientist's model-making function, Feynman selects J. C. Maxwell's work on electromagnetic effects. Maxwell discussed his ideas in terms of a model in which the vacuum was like an elastic solid. He also tried to explain the new equations he got—Maxwell's equations (see section *Einstein's mind* in Chapter 1)— in terms of a mechanical model. Feynman writes:

'Today we understand that what counts are the equations themselves and not the model used to get them. We may only question whether the equations are true or false. This is answered by doing experiments, and untold numbers of experiments have confirmed Maxwell's equations. If we take away the scaffolding he used to build it, we find that Maxwell's beautiful edifice stands on its own. He brought together all of the laws of electricity and magnetism and made one complete and beautiful theory.'

Style in scientific papers: analyst or constructivist?

Since Newton's time, scientific papers have always been couched in positively seigneurial language. Compared with this entrenched manner of address, dry and formal, the more anecdotal, informal mode of communication appears untidy and carries with it less intellectual weight. A second way of writing—the so-called synthetic approach— derives from the style of the great French mathematician and philosopher, René Descartes. Unlike the analytical style, it is not cut-and-dried, and it seeks to persuade rather than to bludgeon the reader into agreement; by so doing, the synthetic presentation invites in its audience a critical frame of mind. In his papers, Descartes preferred to describe his scientific and mathematical discoveries as tentative findings only— like Topsy 'they just grow'd'—and he presented the steps leading up to the discovery in the very order in which they really happened. His successor, Newton, on the other hand, presented his papers as immutable truths, which both awed his reader into admiring acquiescence and inhibited any come-back. Where Descartes carried his reader along with him in an anecdotal style that stimulated critical imagination, Newton advanced his case magisterially in a 'but me no buts' style that has, perhaps unfortunately, stuck—'unfortunately' because Newton's analytical style is that much harder to follow. This may be one of the reasons for the popularity of such a style with scientists: as a device to forestall criticism. And thereby hangs a tale.

As Keynes wrote in his brilliant paper 'Newton the Man', prepared for Newton's tercentenary celebration in 1946, Newton was 'profoundly neurotic'. He had a morbid fear of criticism and a neurotic compulsion never to reveal his thoughts—attitudes which proved to be perfectly justified, to judge by the contumely heaped on him by his scientific contemporaries when first he published his theories of light. K. F. Gauss, too, was secretive about his discoveries, but rather for reasons of prudence: one wouldn't invite anybody to view 'a cathedral with the scaffolding still up' was his reason for tardy publication. So it was for good reason that Newton adopted a style of 'glacial remoteness' for writing his *Principia*. He concedes this fact in writing:

'Upon this subject I had, indeed, composed the third book in a popular method, that it might be read by many, but afterwards . . . to prevent the disputes which might be raised upon such accounts, I chose to reduce the substance of this Book into the form of Propositions [in the mathematical way], which should be read by those who had first made themselves masters of the principles established in the preceding Books.'

A heart-warming exception to the analytical style is the writing of physicist Ernest Rutherford, who is well known for splitting the atom for the first time. In a paper on the effects of oil on water's surface tension (the pull of the water's skin) he recalled a conversation held with a 'Mrs Smith', who apparently travelled with him aboard a steam train to Cambridge and in conversation provided a clue which helped in Rutherford's scientific quest. In true Cartesian style he acknowledged his debt to this chance encounter in a serious scientific paper.

Microbiologists Crick and Watson startled their scientific fraternity by opening their Nobel prize-winning paper with the laconic, Cartesian words: 'We wish to suggest a structure for the salt of deoxyribose nucleic acid (DNA). The structure has novel features which are of considerable interest.'

A similar dichotomy, between aloof analysis and earthy synthesis, colours the teaching of mathematical ideas. On the one hand there is the analytical approach, as taught in most schools, and on the other, the constructivist's approach, limping into some educational curricula. Mathematics, taught from a constructivist viewpoint, would probably become accessible to more people than can be reached by the analytical approach. Indeed, mathematics taught in this way might exemplify Jerome Bruner's aphorism that 'any subject can be taught effectively in some intellectually honest form to any child at any stage of development'.

Newton—master model-maker

In 1964, a new personality appeared on the world's television screens. With quick-fire delivery, a Danny-Kaye wit, Nobel-prizeman and physicist Richard Feynman revealed arcane secrets of physical laws and enchanted viewers with his off-beat handling of mathematical physics, normally the preserve of the super-intelligent alone. The following treatment was inspired by Feynman's dazzling exposition.

When Newton disclosed the awesome truth about the universe, he had to do it with mathematics. No other tool would have done. He expressed, the reader may recall, the force of gravity with which any object, the Moon or an apple, pulls another by the simple equation:

$$F = G\frac{mm'}{r^2}$$

This was the greatest mathematical model ever created. It states that the gravity force F between the two objects (more strictly, the masses m and m') is proportional to the product of the masses multiplied together; the proportionality constant is G, Newton's universal constant of gravity. By this he meant that the force was just as strong anywhere in the universe and mysteriously 'acted at a distance' with no intervening medium or agency. In his day, his theory was regarded as little short of heresy and, at best, wildly provocative. But then a model is nothing if not provocative!

The formula also says something else: the pull between the masses— and it is a *pull*, for gravity only attracts and never repels—is diluted the greater the distance r between the two masses. In practical terms, a spacecraft 10 000 kilometres from the Earth's centre will be pulled down four times as strongly as a spacecraft 20 000 kilometres away, that is, twice as far away. (Newton was careful to measure all distances from the centres of the Earth, the Moon and his celebrated apple, not from their 'skins'. The physics of the situation demands this for technical reasons we need not go into here.)

Why was this model so astonishing? Why was Newton's 'inverse square law' the most exciting law ever formulated? In Newton's day few scientists accepted Galileo's view that a body continues moving in a straight line as its natural state; most clung to Aristotle's belief that the ideal state was one of complete rest and stillness such as they believed the Earth to enjoy. A fashionable seventeenth-century theory had it that angels pushed the planets round in their orbits as otherwise the planets would have stopped dead in their tracks. Perhaps this seemed no more outrageous a notion than Newton's keystone idea of action at a distance. Add to this picture Aristotle's extraordinary law— that no projectile could have two motions at the same time (Renaissance gunners thought that a cannon-ball fired into the air hurtled straight as the proverbial arrow, then jack-knifed suddenly, and plummeted vertically Earthwards)—and the reader has some idea of the prevailing intellectual fog through which Galileo and Newton had to grope towards the scientific light. Shakespeare reflected the current view of the universe when he had Hamlet write to Ophelia:

> 'Doubt thou the stars are fire;
>   Doubt that the sun doth move;
> Doubt truth to be a liar;
>   But never doubt I love.'

It was commonly held that only the Sun moved, and not the Earth; but Galileo reputedly muttered *sotto voce*, after his trial and public recantation, 'Eppur si muove' ('I still say it moves!'). Of course, we now know that both move—at fantastic speeds.

To complete the picture, we must remember that Galileo's and Newton's scientific contemporaries proudly disdained to do experiments in the Baconian manner; only pure thought was considered worthy by these armchair scientists, who actually criticized Newton's first work, *Optiks*, because he had stooped to conquer nature's laws by experiment. Newton had hoped he could convince his colleagues that nobody can discover the truth merely by thinking about it. He failed, understandably shrinking from the biting attacks of his scientist-critics, and devoted his time to discovering why the planets, Moon, and Sun move as Kepler and Galileo had shown they did.

One moonlit night in 1666 the legendary apple fell, though not, as popularly thought, on Newton's skull. But its impact on him was no less great, for it released a creative spark that catalysed and transformed his early, prolonged thinking. The jigsaw pieces of his mathematical model suddenly clicked together. We may fondly imagine that the apple jogged his mind into making his mental leap to the Moon and, as a result, showing that it fell in precisely the same way as the apple. The Earth's gravity pulled it as it fell through the night sky, just as surely as it pulled the apple. Moon and apple, each was slave to the universal tug of gravity. We must also remember that Newton had never seen the 'pull of gravity' as an isolated experiment—as perhaps many people today have not. The authors recall their own sense of wonder at seeing two suspended, leaden spheres in a physics laboratory swing almost imperceptibly towards fixed spheres under the all-pervasive pull of gravity.

True to his preaching, Newton tested his mathematical model. He knew the Moon's orbit to be nearly a circle, with a radius equal to its distance from the Earth—about 240 000 miles. The Moon is tugged by the Earth's gravity, very much weakened by the time it reaches the Moon. The apple falls 16 feet in one second, as Galileo had shown. So the Moon, Newton reasoned, should fall inward towards the Earth that much less in a second because the pull is that much weaker. The reader may like to check the figures; a race-goer's reckoning and some school trigonometry are sufficient for the task.

How far does the Moon fall into the Earth in a second? The Moon takes 28 days to orbit the Earth, so we reckon how far the circle of the Moon's orbit has fallen below the straight line it would have taken if it had not been pulled into the Earth. Figure 43 shows that the Moon falls a distance x centimetres.

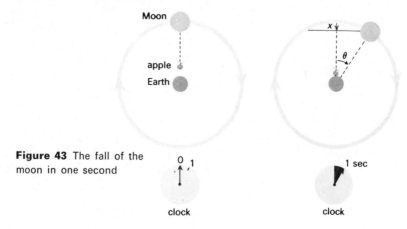

Figure 43 The fall of the moon in one second

$$x \simeq 2R\left(\frac{\theta}{2}\right)^2$$

$$= \frac{R\theta^2}{2}$$

When we put in the values for R, the distance of the Moon from the Earth, and for the angle θ the Moon sweeps through each second, x turns out to be about a millimetre. That much was not new in Newton's day. Now comes the use of his mathematical model.

The apple is 4000 miles from the Earth's centre, and the Moon about 240 000 miles. So the Moon is sixty times as far away from the Earth's centre as the apple. So by the time it gets to the Moon the pull of gravity has been weakened by 60×60 (the *square* of $60 = 3600$) because of the inverse *square* law. In this case, the Moon should fall 1/3600th the distance fallen by the apple in a second—that is, 1/3600 of 16 feet, which is about 1 millimetre. Dramatic evidence that Newton's model worked!

Naturally, Newton did not make his discovery as simply as that. As it was, the then-accepted value he used for the Earth's radius was 15% too small. So his calculations, performed much as we have sketched, did not tally with other known facts. A revised, more accurate value that he used years later showed, however, that his reasoning was impeccable. His own recluseness coupled with the scars of bitter altercations between him and the scientist, Robert Hooke, combined to make Newton stow his original, inaccurate calculations in a table drawer for years, safe from possible ridicule.

What extraordinary repercussions resulted from the inaccurate value of the Earth's radius! Not only was Newton led to distrust his own model*; two centuries earlier, Columbus had mistaken his landfall on America for one on China. It had been a simple matter for even the ancients to measure their latitude, North or South of the Equator— simply by 'shooting' the Pole Star, which is always seen straight up above (at an angle of 90°) at the North Pole, and level with the ground (that is, 0°) on the Equator. But it was quite another thing to measure East–West longitude accurately. And without such know-how, navigation was unreliable. In the section *The problem of longitude* in Chapter 7

---

* According to some commentators, Newton's delay in publishing his theory of universal gravitation was because of the difficulty in proving that the gravity-pull on a body acts as if all its mass were shrunk to the centre. This proof cost Newton enormous intellectual effort.

we see just how this problem was overcome by the great clock-maker, John Harrison. His highly practical work in effect set the stamp of approval on Newton's abstract model. However, in Harrison's day a researcher—like a latter-day inventor—had to prove the value of his work before he was paid; nowadays a scientist is paid while he is doing his research.

A mechanist's dream universe

Newton's model of the universe underscores an extraordinary fact that only a mathematical model could so reveal For mathematics is not simply another language, like a more abstruse literary French or a more elevated *hoch Deutsch*; it is a language plus reasoning. It links one statement with another by an unbreakable chain of logical links, and in this respect it is distinguished from other forms of language. To put it another way, as a mathematician known to the authors quipped: 'A mathematician doesn't think differently from anyone else—just *more*!' Arguably, a matter of degree rather than kind!

We might try to avoid Newton's abstruse mathematics by suggesting some mechanism behind the planets' motions. We might think of the Earth and the Moon hurtling through a cloud of tiny grapeshot, incessantly peppered from all directions. The Moon shields the Earth and Earth the Moon from grapeshot along the direct line joining them. So the overall effect will be that grapeshot from outer space will bombard them more on their far sides than their near sides, and this will gently nudge them together all the time. It so works out that the area each body shields on the other is as the inverse of the square of the distance between them. Again we have an inverse square law, but without paying the price of introducing mathematics . . . or so it seems until we start to generalize the grapeshot mechanism. We only have to consider the Moon on its own, orbiting the Earth, for our model to break down. Just as a car speeding through rain catches more raindrops on the windscreen than on the back window, so would the Moon thud into more grapeshot on its leading face than on its trailing face. This means the Moon would be slowed down and soon come to a standstill. This mechanistic model is not equal to the task of binding together every one of the enormous complexities of nature, with all its amazing laws and rules. The mechanist might be described as one who unconsciously feels that everything and everybody are mechanistic—except himself! 'If you do not appreciate the mathematics,' Feynman once said, 'you cannot see among the great variety of facts, that logic permits you to go from one to the other.' Described fancifully, mathematics is an inter-village helicopter service linking villages of concepts set in impassable mountains.

An armchair Keplerian model

Voltaire was delighted by Newton, who stayed at home and yet encompassed the Earth better than the Paris astronomers who travelled to Lapland and to Peru with their elaborate instruments. Johann Kepler, as we know, whetted Newton's intellectual appetite by calculating the planets' motions about the Sun. Kepler's workings of Tycho Brahe's astronomical observations revealed that the planets moved in ellipses round the Sun; it was as if each planet was chained to the Sun—each a celestial Cerberus on an elastic leash, which was swept round by its ever-racing 'dog planet'. As Kepler found, the 'leash', technically a *radius arm*, sweeps out the same area every second. Figure 47 illustrates this. As the planet swings round the sharpish bend at the far tip of the ellipse it can afford to dawdle a bit after its swift dash down the straighter part of its elliptical course. It makes up for lost area by now having a longer radius arm. On the other hand, at the other tip of the ellipse, nearer to the Sun, the planet has to pick up speed to sweep out its ration of area in the allotted time.

The beauty of Newton's model is that, simply by supposing that the Sun continuously tugs the planet towards itself on its celestial leash, Kepler's 'equal area' result follows quite logically. Here in modern notation, but otherwise unchanged, is the demonstration Newton gave in his *Principia*. In fact, only mathematics can trace the path from Newton's central force model to Kepler's equal areas result. Let us think of the Earth streaking past the Sun. First, we may pretend there is *no* such central force pulling the Earth towards the Sun. As Galileo was the first to discover, the Earth would fly past, straight as an arrow, as in Figure 44. What areas would its headlong course sweep out?

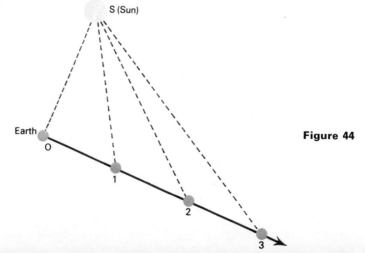

Figure 44

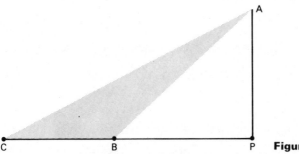

Figure 45

At this point we should remind ourselves of some geometry of the triangle, perhaps long-forgotten. In Figure 45, the height of the peak of the triangle ABC above its base-line is AP. The triangle's area (shaded) is obtained by multiplying 'half the base times the height' ($\frac{1}{2} \times$ BC $\times$ AP). To return to the sweep-area problem, the Earth speeds along the distance 1 to 2 in the same time as from 2 to 3 (see Figure 44). Both triangular areas swept out have the same height O to the Sun or OS. Thus the triangles 1S2 and 2S3 have the same area (same height and equal base). This means that the Earth sweeps out equal areas in equal times, as Kepler declared.

Evidently, then, with no force, the Earth sweeps out equal areas. But what does it do with what Newton in his *Principia* called a central force, pulling it in to the Sun? His argument relied on nothing more fearsome than the area property of triangles and a simple-seeming but really subtle bit of physics. This time Newton envisaged the Earth as streaking past the Sun in a bee-line from position 1 to 2 to 3 (see Figure 46). Then came the 'Catch 22' in his argument. The planet's average

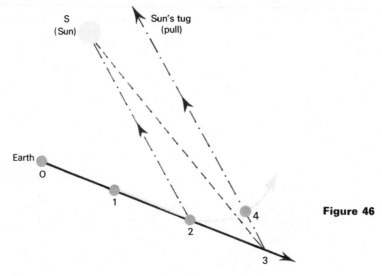

Figure 46

position is at 2. He supposed the Sun-force, which pulls the Earth off its bee-line over to position 4, to act *on average* in the direction 2 to S. So the swerve takes place on the second leg of the journey only. The Earth glides straight as an arrow from 1 to 2 when something happens: it is suddenly tugged steadily by a force acting along a line from 2 to S, the Sun. Instead of reaching 3, it is pulled off-course to 4. Here we should note the subtlety: the line 34 is parallel to 2S, the line of the Sun-pull. What about the areas of the triangles 2S3 and 2S4? Both have the same base 2S, and both have the same height *h,* the distance between the parallel lines 2S and 34 extended. Almost mysteriously, we reach the very same conclusion: the Earth's radial arm to the Sun still sweeps out equal areas.

Mathematics has whisked us on a magic carpet from Newton's central Sun-force model to Kepler's equal areas property. To the mechanist's mind, this feat of abstraction rests on something more sinister. Newton's central force implies the dreaded 'action at a distance', so disturbing to his contemporaries, who far preferred to picture angels pushing the planets round their orbits from behind. Nothing wrong with their picture really except that the angels were pushing in the wrong direction—they should have been nudging the planets in towards the Sun! We do not propose to advance the rival theories, but the problem still remains: 'What is it that acts at a distance? How fast does the agency of this force move?' Modern research on gravity sheds a glimmer on this so far unexplained phenomenon by focussing on the *graviton,* a bullet-like packet of gravity waves (analogous to the quantum of light called a *photon*). Of course, this may strike us as no more and no less acceptable than action at a distance.

Goose-chase after a super-model

It is hard to believe that even today the proper use of models is not fully and widely appreciated. As the physicist is painfully aware, it is possible to picture an electron on Mondays, Wednesdays, and Fridays as a bullet-like particle, and on the other weekdays as a wave, while, as one wag put it, 'on Sundays, it is indeterminate'. The physicist would never dream of treating an electron as a particle in a situation that called for wave-like behaviour. He has had to learn to live with the schizoid duality shown by the electron. This duality has, incidentally, an illustrious history dating back to Newton, who backed the less profitable 'particle' theory to explain light's quirky behaviour. Newton's deadly rival, the Dutchman, Christian Huyghens, plumped for a wave-like picture of light. Today scientists know light plays a double role, just like the Janus-faced electron.

Bitter disputes arose round the rival wave and particle models—reputations were made and lost, and even now, the acrimony long forgotten, the lesson about model-making has still not been taken to heart. The fact is, scientists still seek an all-embracing model of the universe, as Einstein did, fruitlessly, for the last forty years of his life. Hermann Bondi once pointed out the impossibility of ever achieving such an ultimate theory. This goose-chase after a super-model amounts to a hunt for an intellectual Eldorado. Worse, perhaps, is the tendency to accept rival models without bothering to check whether they conflict or are compatible. A case in point is the blind acceptance of the apparently conflicting Galilean and Newtonian models for the trajectories of all Earth-bound bodies, such as cricket-balls, or droplets in a fountain or waterfall. Galileo correctly pointed out that all these follow parabola-shaped paths, whereas according to Newton's book, bodies travelling in space trace out ellipse-shaped paths in the sky. Now there is no reason to doubt the validity of either model. Man's faith in Newton's model borders on the religious. More money and ingenuity have been expended on its proof—by landing men on the Moon—than on any other theory or, probably, any belief. The mathematicians are truly the unsung heroes of the space-age drama of manning the Moon.

Galileo's and Newton's models are not, of course, incompatible. Galileo's is merely a particular case of Newton's wider scope model. The mathematics—which the reader may skip if he likes—works out as follows:

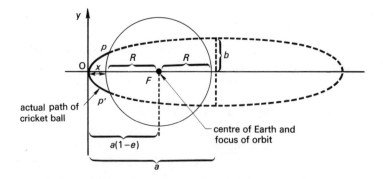

Figure 47 The elliptical path of a cricket ball

The cricket-ball's path through the air is shown by the thick line. It is part, as Kepler pointed out, of an ellipse, with one focus F at the Earth's centre. (An ellipse, has two foci: the other is to the right of F, way out in space.) Now, suppose the ball could pass clean through the Earth. It would complete one elliptical orbit, shown by the dotted line. Of course, the real path of the ball (thick line) is but the merest tip of the ellipse. So the height of the cricket-ball's path above the Earth—we call it x—is fantastically small compared with the full tip-to-tip distance of the ellipse. Half the tip-to-tip distance is labelled a. Evidently, a is as great as (if not greater than) the Earth's radius R. In short, x is tiny compared to a. Or mathematically, $\dfrac{x}{a}$ is small, so $\dfrac{x}{a} \times \dfrac{x}{a}$ or $\left(\dfrac{x}{a}\right)^2$ is negligible, it is so small. (Check: a tenth is small; a tenth of a tenth—a hundredth—is very small.) The equation of the full ellipse (dotted plus thick line) is:

$$\frac{(x-a)^2}{a^2} + \frac{y^2}{b^2} = 1$$

Doing some algebra we get:

$$\frac{x^2}{a^2} - \frac{2x}{a} + \frac{y^2}{b^2} = 0$$

Ignoring $\dfrac{x^2}{a^2}$, we get:

$$y^2 = 2\frac{b^2}{a} \cdot x \qquad\qquad (1)$$

Students of kinematics will recall the equation of a cricket-ball projected with velocity V and angle of elevation α. Take axes x and y at the

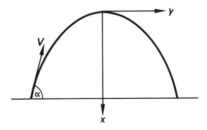

Figure 48

tip of its trajectory along the tangent and the normal to the path as shown (Figure 48). Then the equation given in the textbooks is:

$$y^2 = \frac{2V^2 \cos^2 \alpha}{g}.x \qquad (2)$$

Equation (2) has precisely the same form as equation (1), that is,

$$y^2 = K.x$$

So we have shown that the Galilean parabola model for cricket-balls and the like is really a special instance of the larger Kepler–Newtonian model for spacecraft, planets and satellites—and for cricket-balls that can burrow through the Earth!

For instance the equation of the orbit of Russia's *Cosmos* satellite (see Figure 49) is:

$$\frac{(x - 4455)^2}{4620^2} + \frac{y^2}{4600^2} = 1$$

And at the very tip of this orbit, the path is pretty well a parabola.

Armchair physics

To dispel a feeling of being 'had' by mathematical sleight-of-brain, we can establish the same picture by physics. We begin with the principles of the conservation of energy and of momentum—twin cornerstones of the physical sciences. From them can be derived a mathematical expression for the shape of the orbit of a projectile—be it an apple or Apollo. Called the *eccentricity*, this expression is a measure of an oval's roundness or elongation. A perfectly round billiard-ball, for instance, has an eccentricity of 0; the oval of a rugger ball has a value of $\frac{2}{3}$. The longer the oval, the nearer the value approaches 1, which is the value for a parabola—an infinitely long oval, in effect. The expression for the eccentricity (e) in terms of the projectile's velocity parallel to the Earth's surface is really very simple:

$$e = 1 - \frac{v^2}{V^2}$$

As before, v is the projectile's velocity. And V is the launch velocity which will just put it into orbit round the Earth (that is, about 18 000 m.p.h. or 29 000 k.p.h.). For a bowled cricket-ball or fired rifle bullet, neither of which go into orbit, v is so much smaller than V that v^2/V^2 is practically zero, so $e = 1$, the known criterion for a parabola. This result happily agrees with Galileo's model! For *Apollo*, with v equal to the launch velocity V, we have $e = 1 - V^2/V^2 = 1 - 1 = 0$, the criterion of a circle, as predicted on Newton's model! A typical space orbit—for Russia's *Cosmos* satellite—is compared with our cricket-ball's imaginary flight through the Earth (see Figure 49). The satellite orbit's eccentricity is 1/10, very nearly the criterion for a circle (which is 0).

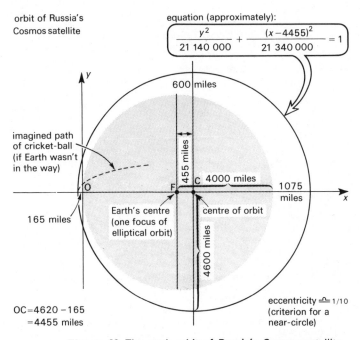

orbit of Russia's Cosmos satellite

equation (approximately):
$$\frac{y^2}{21\,140\,000} + \frac{(x-4455)^2}{21\,340\,000} = 1$$

600 miles

imagined path of cricket-ball (if Earth wasn't in the way)

455 miles

C 4000 miles

1075 miles

165 miles

Earth's centre (one focus of elliptical orbit)

centre of orbit

4600 miles

OC = 4620 − 165
= 4455 miles

eccentricity $\doteq$ 1/10 (criterion for a near-circle)

Figure 49 The earth orbit of Russia's *Cosmos* satellite

Finally, a modern comment on model-making in physics. As late as 1955 nobody had found the reason for not using pendulum clocks in testing Einstein's model of the universe. It was well known that many kinds of clocks can be used for measuring time over space journeys, which feature in the so-called Einstein-clock paradox. Bondi recommends rabbits, decaying radioactive particles, or atomic clocks for the job. But what must *not* be used, he points out, are pendulums. The German physicist, Max von Laue, explained why: a pendulum is not just a toy in a box bought from a shop. In the 'package', the shopkeeper is actually selling the Earth as well! And, in the words of the old song, you can't take it with you. For instance, an Earthly seconds pendulum clock on the Moon would beat about $2\frac{1}{2}$ seconds.

A model of colour vision

Colour vision has always appealed to physicists and mathematicians. So thought Feynman, and the following discussion is adapted from his intriguing ideas.

The great quantum physicist, Erwin Shrödinger, developed a wonderful mathematical theory of the mixing of colours, which was based on *vectors*. In concise terms, a vector can be described as a mathematical 'arrow' which has length and a direction. A jet speeding at 1000 kilometres an hour in a north-westerly direction (that is, 45° round the compass from North) has a certain velocity, and the mathematician tags its line of movement with the two numbers (1000, 45)—a vector. As for a code, the first number gives the ground-speed, and the second number the direction in degrees round the compass. We can write vectors with far more numbers in them—and they need not be obvious physical quantities. A cinema manager, for example, might record his day's takings as the vector (100, 250, 300, 150)—each number being the number of seats sold in each of the four price ranges.

Two features of vectors are important. A vector can be thought of as an array of pigeon-holes; and vectors are composed uniquely. As long as the numbers remain within their own pigeon-holes, then two vectors may be added in 'schoolroom' fashion. So the cinema manager can add the first takings with another set (0, 10, 50, 100), taken on a bad night! He adds the corresponding numbers within pigeon-holes—100 to 0, 250 to 10, and so on for both nights' box office taking. The total takings are then given by the vector (100, 260, 350, 250). The second feature means that the pigeon-holes never overlap. The cinema manager does not sell tickets that would allow a person to hover between two differently priced seats throughout the show! Technically speaking, the vectors are linearly independent.

Now that these two features of vectors have been established, let us try to put colour vision on a mathematical footing. Psycho-physical experiments with coloured lights and the eye reveal two great laws of colour mixing. These are modelled exactly on the two fundamental rules of vectors—concerning addition and 'self-contained' components.

The first colour law says that we can make various colours in more than one way by mixing lights from various filters. (To return to our cinema manager for a moment, he could have the same gross takings on *different* nights from different combinations of seats sold.) We can put this analytically. Yellow ochre, say, denoted by the symbol Y, is the 'sum' of specified amounts of red-filtered light (R) and green-filtered light (G). The brightness of the red light is r, say, and of green light g. So we can write a formula:

$$Y = rR + gG$$

In words, yellow ochre is a mix of a fairly bright red and a not-so-bright green, perhaps.

Now for the second law of colour mixing. Suppose we match two shades of yellow light, indistinguishable to the eye, with different proportions of red and green and some blue in it. We could write this as $Y = X$. Here X is some other mix of colours that produces an effect of yellow ochre Y. If we now add a light pink (P), for instance, to each of the ochre lights Y and X, they will still match! In algebraic terms we have:

$$P + Y = P + X$$

And this equation balances just as well as $Y = X$.

This law is modelled on the law governing the addition of mathematical vectors. It actually works even when the eye is colour fatigued. When we stare at a bright red surface and look away at a white wall, the wall looks green. Suppose now that we look at two matched yellows Y and X, say, against a brightly lit surface and then look at a white wall. They will not still look yellow. But they will still match! We have added some colour Z to both tints X and Y but they still match, like this:

$$X + Z = Y + Z$$

(The only time the law breaks down is in very dim light when a different part of the retina of the eye is operating—in fact, the rods instead of the cones.)

A third law says: any colour can be made from three different colours, say red (R), green (G), and blue (B). So any tint T can be expressed as:

$$T = gG + rR + bB$$

where blue-filtered light has a brightness b. Another colour C can be made out of a different mix of the same primaries, thus:

$$C = g'G + r'R + b'B$$

To get a mixture of the two lights, we take the components (the amounts of the primaries in each colour T and C) and add them—exactly as for vectors. So a mixed colour (M) of T and C is given by:

$$M = T + C = (g + g')G + (r + r')R + (b + b')B$$

Ask a mathematician to add two vectors T and C, with the same components, (g, r, b) for one vector and (g', r', b') for the other vector, and he will arrive at the last formula for M—because vector elements are linearly independent.

Of course, we may not be able to get any colour we want without doing something more. We cannot match a green light, for example, with a mixture of the three primaries. But, shine some red light on the green patch of light, and we *can* match the patch with such a mixture. In terms of algebra, this means we must add something to both sides of the mixing equation. That is to say, we may have to add a negative amount to the right-hand side of the equation. This again tallies with the vector operations allowable by a mathematician.

Tolerance and uncertainty

Here follow some thoughts of a philosophical turn, which the reader may skip if he wishes, without fear of losing the thread of the authors' theme. It has been stated in the preceding section that one yellow ochre (X) is the same as another (Y)—that is, $X = Y$. Perhaps it would have been nearer the mark to say that the one ochre is indistinguishable from the other; any difference is undetectable by the eye—or, expressed in symbols, $X \sim Y$. True, this spoils the nice matching of colour vision with vector analysis—isomorphism of structures, in fact, exists in both ideas—yet it comes nearer the truth. It breaks the umbilical cord that all but physicists hang on to, a lifeline belief that the world, like the realm of Euclidean geometry or the differential calculus, is a place of smooth, imperceptible changes and graduations.* The fact that the two colours are indistinguishable does not necessarily mean they are exactly the same. The hues are, after all, mixed differently.

This marginal note has striking repercussions, as we shall see in our look at a model of the brain (see section *A model of the brain* in Chapter 5). What is being suggested here is that the two yellows, X and Y, are near enough the same thing; there is a tolerance between them, as there inevitably is between, say, one pair of size-10 shoes and the next pair.

* Of course, physiologists are well aware of all-or-none discrete jumps in neural states.

As on an everyday level, where nothing is neat and tidy and where we have to make allowances, so it is on an atomic level. Physicists have had to learn to live with a rule book in which the cardinal law reads something like Bernard Shaw's 'You never can tell from where you sit', more formally known as the 'uncertainty principle'. This principle says, in so many words, that if we peer at an electron and in doing so bombard it with even smaller particles, the massless photons of which light is composed, we affect the electron's behaviour. Fancifully, we may imagine the electron as being shy in front of an audience of scientific observers. The net result is this: we can never tell exactly where an electron is—it may be one of the electrons peppering the screen on our telly tube, for instance—nor how fast it is going. We can, however, know its velocity and its position pretty exactly, that is, within a certain tolerance or margin of uncertainty. We can put it in mathematical terms; first we experimentally determine the electron's position and its velocity within a certain margin of error. When we multiply the errors in the position and the velocity together we get Planck's constant, h, a very small quantity indeed. If we insist on guiding an electron in the TV tube to dead centre of the screen, so that we can pin-point where it is, then we discover that it strikes the screen at an indeterminate speed—known only within a range of values. But force it to go at a precise velocity, and its possible position 'spreads out' like a vague cloud. This fuzziness might be characterized as the mediaeval philosopher's 'cloud of unknowing'! It troubled Einstein, for one, so much that he was once heard to say, 'But God does not determine how electrons should go by throwing dice!'

The fact is, this aleatory chanciness associated with physics is now accepted by all but a few scientists—when they are being scientists. But a curious change takes place once they form mathematical models. Enunciating his famous *conventionalism concerning geometry*, Poincaré once said that 'no empirical test [such as triangulation in making geographical maps which amply bears out the correctness of the congruence of triangles], whatever its outcome, can conclusively invalidate the Euclidean conception of physical space'. Of course, the validity of Euclid's geometry can always be preserved, even at the major expense of changing physical theories. This calls to mind, humorously, lines from Lady Bracknell's famous exchange with her prospective son-in-law, Jack or Ernest, the eponymous hero of Oscar Wilde's *The Importance of Being Earnest*. She is quizzing him about a house he owns in Belgrave Square.

Lady Bracknell: '. . . What number in Belgrave Square?'
Jack: '149.'
Lady Bracknell: (*Shaking her head.*) 'The unfashionable side. I thought there was something. However, that could easily be altered.'
Jack: 'Do you mean the fashion, or the side?'
Lady Bracknell: (*Sternly.*) 'Both, if necessary, I presume.'

To follow Poincaré's recommendations today would mean changing *both* scientific fashion *and* physics. For, ever since the arrival of Planck's quantum theory, space on an atomic level has been unmasked as decidedly non-Euclidean. In physicist Niels Bohr's model of the atom, the electron is believed to jump from one orbit to another without actually existing in between, and this is reminiscent of nothing so much as the conjurer's cry 'Now you see it, now you don't!' The orthodox explanation involves Schrödinger's differential equations, which too rely on a smooth three-dimensional mathematical *ether*.

Readers may recall that at the turn of the last century most physicists believed in the existence of an all-pervading ether, through which all phenomena such as light, heat and radio waves were supposed to permeate. The theory proved untenable, as did the idea of a smoothly continuous mathematical ether for handling atomic phenomena. To quote the topologist Zeeman:

'The root of the trouble lies in the effort to fit the strait jacket of Euclidean space onto the universe. After all, nothing in physics suggests the existence of so sophisticated a mathematical construction as the real numbers, let alone the wildness that can occur in Euclidean 3-space—that is, in our conventional three-dimensional world.

'Nothing', he goes on, 'in physics suggests even non-countability.' It is a well-known, but tiresome fact of Euclidean geometry that there are an infinite (uncountable) number of points between, say, two points on a line. A far cry from the uncuttable atoms of Democritus!

Nevertheless, there persists the belief that physical phenomena come in discrete packets. Zeeman proposes doing away with the Euclidean ether, and adopting laws that work within certain tolerances, finer than which it is inhuman and bootless to go. The obvious tolerance to go for is that invoked by the uncertainty principle, he suggests. There is no earthly reason to posit a theory that grinds space into finer particles than the uncertainty principle will allow us to detect.

The problem has a parallel in the consideration of time and motion, on which the Eleatic Zeno of ancient Greece focussed so tantalizingly in his celebrated paradoxes. The most famous, Achilles and the tortoise, which is the first recorded discussion of a *limit*, the linchpin of the calculus, is elegantly described by Tolstoy in his *War and Peace*:

'The human intellect is incapable of understanding *a priori* the idea of unceasing movement in a body; it can only apprehend it when it is at leisure to analyse the component factors and study them separately; at the same time, it is this subdivision into definite units which gives rise to many errors. For instance, a well-known sophism of the ancients tended to prove that Achilles could never overtake a tortoise crawling in front of him even though he walked ten times as fast as the animal; for, every time Achilles should have picked up the distance between them, the tortoise would have got ahead by a tenth of the space; and when Achilles had covered that tenth the tortoise would again have gained a hundredth, and so on, *ad infinitum*. The ancients regarded this as an unanswerable dilemma; its absurdity lies in the fact that the progress of Achilles and the tortoise is calculated on diminishing units with stoppage between, while it is in fact continuous.

'By assuming the minutest units of any given motion as a basis of calculation, we may constantly approach a solution without ever reaching it; it is only by admitting infinitesimal quantities and their progression up to a tenth, and adopting the total of this geometrical progression, that we can attain the desired result. The modern science of the value of the infinitesimal solves questions which of old were regarded as insoluble. By admitting these infinitesimals it restores motion to its primary condition of inherent perpetuity, and so corrects the errors which the human mind is led to commit by regarding the separate units of motion instead of motion as a whole.

'In our search for the laws of history the same rule must be observed. The onward march of humanity, while it is the sum total of an infinite multitude of individual wills, is nevertheless uninterrupted; the study of these laws is the object of history . . .'

A model for publishing

Descending from the science-sublime to the business-mundane, light relief is offered in the form of a mathematical model designed by the authors for the world of publishing. Not only is the model a novel application of the calculus—even, perhaps to publishers—but it is a tenable exemplification of the craft of 'numeracy'. The Crowther Report, which coined this word, missed half the point when it championed the need 'in the modern world to think quantitatively, to realize how far our problems are problems of degree even when they appear as problems of kind'. No business man would be in business if he did not quantify. But he has rarely thought at one stage removed, in the abstract, that is, like a mathematician. We must not forget the mathematician's legendary ineptitude in handling men—perhaps a reflection of his typically introverted personality—which makes him so unsuitable for the responsibilities of the manager's job. Lord Snow put his finger on the touchstone of a good leader-manager when he wrote, 'The lesson to the scientists'—and he might have included the mathematicians—'was that the prerequisite of sound military advice is that the giver must convince himself that, if he were responsible for action, he would himself act so. It is a difficult lesson to learn.'

And this is the lesson any private publisher—alas, a practically extinct business species—or any entrepreneur must learn or perish. Typically, the manager has regarded 'people' as the problem which he is especially suited to handle; in any enterprise his is the 'judgement' component, with the (to him) lesser 'quantity' element of the business left in the capable, but inferior hands of another—often a mathematician. Worse still—and here the relevance to 'numeracy' is paramount—as management consultant Albert Battersby once wrote, the manager may not 'be prepared to sacrifice the clutter of routine figurework and to accept that more of his job may be reducible to mathematical procedures than his pride cares to admit'.

Any reader may be forgiven if he gets the impression that a clutter of figure-work, correctly executed, epitomizes the Crowtherian notion of 'numeracy'. If it were, it would be a sorry educational aspiration. An example drawn from the experience of one of the authors as an editor in a publishing house may dispel this false impression.

It is the custom in a book publishing house to cost a book, at least one for the educational market, according to a definite pattern. And it is also customary for costings to be reworked each and every time one of the factors affecting the book's cost is changed. For instance, a publisher proposing to sell a softback at 30p will have to contend with a number of ponderables—trade discount, author's royalty, and overheads and gross profits necessary for him to keep alive. On the basis of figures for each of these 'parameters', as they are called, the publisher's production manager can estimate the retail selling price within a fraction of a penny. But if one of the parameters changes—maybe the author demands a higher royalty and the publisher, naturally anxious to accommodate him, accedes, or production costs rocket due to a rise in the cost of paper (a contingency any publisher has to allow for between the production phase and publication date), then the publisher must cut his book according to his paper. For each such change of parameter, the production manager traditionally started his calculations afresh and recosted the book. His zealous industry recalls the Crowther notion of 'numeracy' but was scarcely very good mathematics. To catch something of the new numeracy, we will look at a hard and fast example.

A proposed book's retail price is planned to be 30p. The author is offered a $7\frac{1}{2}$% royalty, based on the retail price, but he pleads for a higher percentage, 10%, say. He argues, plausibly enough, that $2\frac{1}{2}$% of 30p amounts to 0.75p, which is all, he maintains, it would add on to the book's retail price. This, as any traditionally trained business man—numerate or not—will be quick to see, represents an author's royalty not of 10% on 30p but 10% on 30.75p. In a business where the ability to shave 0.125p off the cost of a book may be crucial, provided the book sells in large enough numbers, this naïve assumption could spell failure for the publisher. How should the calculation go then?

The publisher might base his costing on some such figures* as these in which each item (b), (d) and (e) is subtracted from the figure above it:

		p
(a)	Published price	30.00
(b)	Trade discount @ 25% of (a)	7.50
(c)	(a) − (b)	22.50
(d)	Royalty @ $7\frac{1}{2}$% of (a)	2.25
		20.25
(e)	Overheads and gross profits, based on 50% of (c)	11.25
	Production costs	9.00

The series of simple 'sums' reduces to the production costs as the fixed, irreducible that the publisher cannot afford to eat into if he is to maintain a high standard of printing, typography, layout, design, and editorial work. He expresses the retail price in terms of this constant factor in his costing formula. He divides the price by the production costs (30/9.00 = 3.33) to get what he calls his mark-up. As a marginal note, mark-ups in educational publishing are, on the whole, much lower (about 3) than those for books for the general trade market

* These figures are in line with those quoted by Fred Timms, *Bookseller*, 15.9.71.

(upwards of 4), with books for technical libraries having the highest (over 6 sometimes). The mark-up, as the name suggests, is a measure of the publisher's expected profits, all other factors being the same. Holding the production costs steady at 9.00, the publisher naturally seeks a way to meet the author's plea for a larger royalty. How does he do it? The best way is to generalize the situation into an algebraic formula.

First, he works all figures based on a cost of 100. Later he adjusts his figures to the actual retail price (s) by dividing by 100 and multiplying by the selling price s. He calls the author's royalty r. This we know is $7\frac{1}{2}\%$ at the outset, but to avoid reworking several sums unnecessarily when he wants to recalculate for a new royalty, he leaves it in as a fraction (r) in the formula:

$$\text{Production costs} = s\left(100 - 25 - 100r - \frac{50}{100} \times 75\right)\Big/100$$
$$= s(0.375 - r)$$

The expression in brackets ($0.375 - r$) is 1 divided by the mark-up (m), in fact.

Now, with a selling price of 30p he can afford to offer his author a royalty r of $7\frac{1}{2}\%$. So putting $r = 7\frac{1}{2}\%$ (that is 0.075) in the formula gives:

$$\text{Production costs} = 0.3s$$

However, in response to the author's heart-rending request, our soft-hearted, hard-headed publisher hefts his royalty to 10%. This means that the new royalty, denoted by r', say, is 0.1. The public is asked to underwrite this act of generosity by paying more for the book. We now calculate this new selling price, called s' for the moment. The production costs must be pegged to the same figure as before. We can write the equation:

$$0.3s = (0.375 - r')s'$$

or

$$0.3s = (0.375 - 0.1)s' = 0.275s' \qquad (A)$$

This has been labelled formula A because we will return to it. The price *rise* is ($s' - s$). Now s is 30. We put that value in formula A to get the new price s':

$$s' = \frac{0.300}{0.275} \times 30 = 32.8$$

So the price rise (32.8 — 30) is slightly under 3p. And this shows that the actual price rise is not 0.75p, as our author fondly imagined, but more like four times as much!

Here the mathematician steps in and generalizes the situation: he asks 'What effect would changing the royalty to *any* value have on the selling price?' He turns to our formula *A*:

$$(0.375 - r)s = (0.375 - r')s'$$

or

$$ks = (0.375 - r - r' + r)s'$$

where $k = (0.375 - r)$, in other words, $1/m$, or 1 over the mark-up. Now the royalty rise he calls $x = r' - r$. So the equation becomes:

$$ks = (k - x)s'$$

or

$$s' = \frac{b}{(k - x)}$$

where he writes b for ks, a fraction of the retail price. Using the calculus, he differentiates—that is, carries out a process for finding the slope of the equation:

$$\frac{ds'}{dx} = \frac{b}{(k - x)^2}$$

Those who have done the calculus may recall that we find something noteworthy when the expression becomes either zero or infinite. Here, it becomes infinite when $x = k$, giving what is known as an asymptote. Figure 50 shows the curve. When the royalty rise is equal to 1 over the

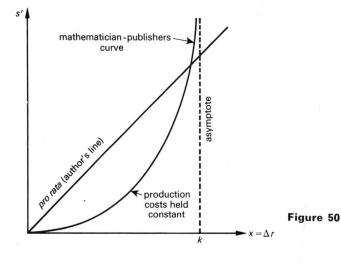

Figure 50

mark-up, the publishing price has to be infinitely high! If the author–publisher could ever achieve this desirable situation even once in a lifetime, his financial problems would be over for good! It is intriguing to compare the two costings, the author's naïve arithmetical *pro rata* assessment with the mathematician-publisher's. The first, a straight line, symbolizes old-style marketing; the second, the up-swept curve, is a swinging mathematical model of business. And when x does equal k, then $r' - r = 0.45 - r$ or both the new and the old royalties are 45%. An unheard-of royalty for any publisher to pay an author!

Drug model

The late and sudden rise to importance of biology as the major growth science is reflected in an upsurge of mathematical applications within biology and medicine. The models developed as a result may eventually profoundly influence medicine and hospital treatment, if not general medical practice. A case in point is the Indian mathematician J. S. Rustogi's model for antibiotic drug levels in the human body. Anybody who has had a course of injections of an antibiotic drug will know that doctors usually prescribe a big, booster dose at the outset of the course. Thereafter smaller regular doses suffice to maintain the prescribed drug level, for the drug disappears from the body steadily. A model of this drug disappearance—a phenomenon of great interest to pharmacologists —has been developed by Rustogi.

In this model, we call the drug level in the human body y at any time t. After a time, the drug will have completely disappeared from the body, and y will be zero, of course. Secondly, we say that the initial drug dose is y_0. The small 0 below the y is a reminder that this is the drug level in the body at the outset, when we set the clock to 'zero hours' as it were.

Making a fairly reasonable assumption, we propose that the greater the amount of drug in the bloodstream, the greater the rate at which the drug drains out of the body. This assumption is crucial. It sets the pattern for this model and for so many others like it. It may help to think of a foyer packed with conference-goers, when suddenly the bar is declared open: then the greater the crush in the foyer (bloodstream) the greater the rush to the bar (draining away from bloodstream). There are, presumably, no drug queues waiting to get out of the blood-stream, and we write the rate of draining away of drug y in a short time *t* as d*y*/d*t*, using the symbolism of the calculus. Since the rate is proportional to the amount of drug present *y*, we write:

$$\frac{dy}{dt} = -ky$$

The minus sign shows that the rate *diminishes* rather than increases, albeit proportional (hence the *k*) to the amount *y*. By integrating our expression—that is, finding what the amount (*y*) is from the known rate (d*y*/d*t*)—we get:

$$dy = -k \int y \, dt$$

or

$$y = y_0 \, e^{-kt}$$

The beautiful symbol $\int$ is the old printer's letter 's', meaning 'the sum of'.

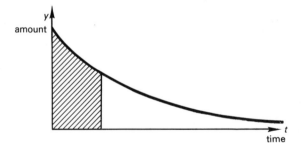

Figure 51

We can picture this draining away of the drug as a curve (see Figure 51). Radio-enthusiasts will instantly recognize this telltale curve as characterizing an electric charge leaking from a condenser. But atomic physicists will claim the model as a picture of the decay of radioactive particles (such as the radium in the luminescent spots on a watch). Chemists, however, will see in it a picture of chemical reaction rates. The exponential model has a protean quality, like that attributed to beauty. In mathematics utility is seemingly in the eye of the scientist-beholder.

If we propose that some two-thirds of the drug disappears from the body in, say, a day, a neat calculation ensues. Our formula then shows that when the drug level has fallen to $1/e$ of its initial value, that is, by about $\frac{2}{3}$ (actually 63.2%), the time elapsed is equal to $1/k$. The e in our formula has a value of about 2.718. After this time $(1/k)$, where k depends on the type of drug, has elapsed the doctor injects another dose of the same size into the system. After a great many doses—mathematically, an infinite number—the drug level becomes:

$$y = y_0 \frac{1}{1 - 1/e} \simeq 1.6 y_0$$

If, instead of giving dose y_0 at the outset, he administers a bigger booster dose y_b of $1.6 y_0$, and suppose $k = e$, then, after a definite time, numerically equal to $1/k$ hours (dependent on the type of drug), the drug level falls to an eth of its initial level—that is, to y_b/e. In that time, $0.6 y_b$ of drug has drained out of the body. So repeated doses of y_0, one dose every $1/k$ hours, bring the drug level back to y_b, the booster level. This means that, with an initial booster dose of a little more than half as much again as the regular doses, each successive dose makes the patient's dose level peak up to the original booster level. A picture of the dose levels is given in Figure 52. The curious feature of this model is that the time between doses is dictated by the kind of drug used. The

Figure 52

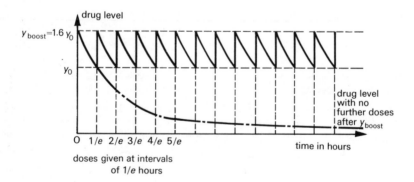

mathematics works out cleanly, as seen in the graph, with a time interval of $1/k$ hours (0.37 hours). It has to be $1/k$ hours, where k is a measure of the speed at which the drug disappears from the body.

The model is reminiscent of the older, Newtonian one for a teacup cooling in a continuous draught. It is as if we were to keep a cup of tea hot by adding dollops of hot water at repeated intervals. Again the same, exponential model fits the facts: the temperature of the tea is proportional to the initial temperature multiplied by e^{-kt}, where t is the time—exactly as for drug levels. Here, the constant k relates to the size of the cup and the cooling conditions, much as k in the drug model related to the actual drug used and the size of the patient's body.

Figure 53

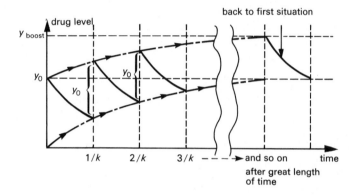

As an interesting sidelight, we can plot a graph to show what happens if the initial dose is still y_0 (with no booster), followed by regular doses of the same amount y_0 (see Figure 53). Curiously, after a great many such doses, the level in the body builds up to a 'saturation' level of $1.6y_0$, that is, y_b, the booster dose level—provided doses are administered every $1/k$ hours.

From Newton's Law of cooling to a drug model may seem a far cry . . . but it's all one in mathematics.

CHAPTER FIVE

The Mathematics of Mind

'How can I know what I think till I see what I say?'
(Alice on being told to think before speaking)
LEWIS CARROLL

'The forceps of our minds are clumsy things and crush the truth a little in the course of taking hold of it.'
H. G. WELLS

'The dream neglects in a most conspicuous manner the logical category of opposition and contradiction. The concept "No" does not seem to exist in the dream. It likes to compress opposites into a unity, or to represent them as one.

The awkward fact—that reason, as we know it, is never aware of its hidden assumptions—has been too much for some philosophers, and even many scientists to admit.'
L. L. WHYTE

'The rules of the game, however absurd, cannot be altered by playing that game.'
ARTHUR KOESTLER

Reductionism versus organicism

Before looking at a mathematician's model of mind, we should survey the contemporary scientific Zeitgeist which informs our thinking. Probably only a poet-mathematician who happens to be a genius could hope to create a model of how we think. In so far as intelligence is the outward show of the mind—as through a glass darkly—we might answer the perennial question 'What is intelligence?' We might define intelligence, in the mathematician R. Thom's words, as 'the ability to make models of the outside world and models of inside thoughts'.

Allied to intelligence is memory, which most neurologists are inclined to believe is linked to the action of the synapses, 'nerve endings', in the brain. The brain has about 10 000 million (10^{10}) brain cells (neurons), each having as many as 100 000 (10^5) synapses—a staggering total. The synapses are the links with the other neurons. When enough synapses fire, then a message can jump from one nerve cell to the next. An interesting fact about the synapses is that, though we can do nothing about the nerve cells—about 15 000 cells die each day until senility sets in—we can keep our minds in trim by thinking, which keeps the synapses open by making them fire. From an educational point of view, every moment in school spent not thinking is tantamount to inviting slow brain death at a time when a person most needs to develop his wits—when he still has a full complement of them—that is, his birthright of synapses. Prolonged mental inactivity leads to atrophy of the synapses, after which they can never be made to fire. When synapses die, due to the sleep of reason, they cannot be resuscitated like Rip Van Winkle. Synapse death is irreversible.

With which sort of functions must a good model of the mind be consonant? What sort of facts must it account for? Probably foremost are the astonishing results of American psychologist Karl Lashley's extirpation experiments on rats' brains; these showed conclusively that memory is not localized in any special part of the cortex. The cortex is the crinkly, wafer-thin layer of grey matter beneath the skull—in form, much like the inside of a walnut—where most of our 'hard thinking' takes place.

In the authors' view, a reductionist, as opposed to an organicist, model will not do. In *Beyond Reductionism* Koestler suggests that many sentient scientists today 'feel a discontent with the Zeitgeist and its reductionist "nothing but" attitude'. And he cites 'four pillars of unwisdom', one of which is 'that mental evolution is the result of nothing but random mutations preserved by natural selection' and another 'that the only scientific method worth that name is quantitative measurement; and, consequently, that complex phenomena must be reduced to simple elements accessible to such treatment, without undue worry whether the specific characteristics of a complex phenomenon, for instance Man, may be lost in the process'. This is the reductionist philosophy in a nutshell, albeit in an unsympathetic one!

In a little-known passage Arthur Eddington assesses the place of entropy in the philosophy of science. He rated it as the greatest contribution to thought by nineteenth-century physics. It marked a radical departure from the previously held view that microscopic dissection of details will reveal all the scientist needs to know. It announced a swing from analysis of the parts to an appreciation of the qualities possessed by a whole system. 'We often think', Eddington writes, 'that when we have completed our study of *one* we know all about *two*, because "two" is "one and one". We forget that we still have to make a study of "and". Secondary physics is the study of "and"—that is to say, of organization.'

More recently the distinguished biologist C. H. Waddington invoked in genetics an indeterminacy—like the physicist's uncertainty principle —to account for the relation between biological form and genetic heritage. Waddington assigns the environment to the role of the unknown, indeterminate factor. To describe evolution adequately, he suggests, simple mathematical treatments such as J. B. S. Haldane and Ronald Fisher applied will not do. Instead, Waddington proposes, it calls for mathematics more like the Theory of Games. (Significantly the Theory of Games, created in the late 1920s, is one of the few branches of the so-called modern mathematics that can claim to be of this century.) Waddington takes this theory to task for being 'promise crammed', as the air was to Hamlet, yet apparently containing almost no theorems. The main one, the Theorem of Minimax Strategy, in effect tells us how to play safe. 'The content of the Theory of Games', Waddington points out, 'seems to be almost non-existent except for discussions of elaborate ways of formulating problems which you then find you can't solve!' and wishes that someone 'would now put some meat into the Theory of Games'. (For what it is that needs 'meat', see the section *Games theory* in Chapter 6 of this book.) Waddington also draws attention to the astonishing way in which computers made up of only a few binary on–off switches settle down into a steady cyclic condition. In short, a randomly built system can be inherently stable of its own accord. The philosopher, Carl Friedrich von Weizsäcker's remarks on reductionism are telling:

'Those reductionists who try to reduce life to physics usually try to reduce it to primitive physics—not to good physics. Good physics is broad enough to contain life, to encompass life in its description since good physics allows a vast field of possible descriptions. There is no reason why living beings should be compared to primitive machines which don't make use of feedback.'

He goes on cautiously and metaphysically to add: 'I see no reason why what we call matter should not be "spirit".' Certainly there is something spiritual in the total belief demanded of a physicist if he is to swallow whole such intangibles as 'action at a distance' or the 'electron jumps' in Bohr's model of the atom.

Bee's-eye view

Before we look at perception and the human brain, it were as well to enquire scantly into a more rudimentary system, the bee's eye, for instance. Like Feynman, we might ask why the bee has not developed as good an eye as our own. What has stopped it from evolving an eye like ours? The short answer is that its eye has to solve different problems from ours and so is bound to be built along different lines. But we can find reasons more compelling if we adopt a simple physical model of the bee's eye, and then carry out a few simple operations in the calculus. (A reader not conversant with the calculus—in a nutshell, the mathematics of very small changes—may admire the beauty of the printed symbols and skip to the conclusion.)

We may think of the bee's eye as made up of myriad lenses like elongated ice-cream cones, with transparent lids instead of balls of ice-cream. All these cones, actually *ommatidia*, like so many contact lenses, are packed to form the surface of a sphere, the outside of the bee's head. Figure 54 shows schematically the packing of the ommatidia in a bee's eye. The sphere, of the bee's head, has a radius of r, and each

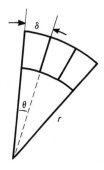

Figure 54 Schematic packing of the ommatidia in a bee's eye

ommatidium is δ across—that is, the diameter of the 'contact lenses', which are all of a size. What happens if the ommatidia are large? Then one ommatidium may see a tree on one side of the garden, and the next-door one may see a flower on the other side of the garden. As a result the hapless (and doubtless haploid) bee cannot see things between. And this is no use, in the evolutionary sense, or any other! So the size of the ommatidium affects the bee's *visual acuity*—whether or not it has 'sharp eyes'. We can measure this size by the *angle* from one ommatidium to the next, that is, the angle of the 'ice-cream cone' itself. This angle, θ say, is related to the diameter δ of the cell and the radius r of the bee's head by the formula:

$$\theta = \frac{\delta}{r}$$

This simply says, in mathematical symbols, that the longer r is, the smaller the cone angle θ; and the larger δ is, the bigger is the cone angle.

So why does the bee not have very fine ommatidia? The physicist can answer for the bee: when light passes through an ultra-fine slot like an ommatidium it may enter the slot from widely different directions. The effect is known as *diffraction*. In short, instead of the cell receiving light head-on only, light from a slightly wider angle A will find its way in. This angle A is linked to the wave-length of light. Like radio, light has definite wave-lengths, red light being the long waves, comparatively speaking, and blue light the short waves. A well-known formula about the diffraction of light links these two physical facts:

$$A = \frac{\lambda}{\delta}$$

where λ is the wave-length of the light.

For the interested reader, the physics of the situation can be best explained by looking at the reverse phenomenon: parallel rays of light pouring out of the bee's eye. (This is how scientists once thought vision operated!) The emergent rays bend and light up a screen in wider and wider circles of alternating rings of bright light and dark shadows. This spreading effect is what we call diffraction—a supreme example of multiple-interference. The mechanism can be modelled very well with two long strips of corrugated cardboard: one is arranged to represent the wave CP, the other BP (see Figure 55(a)). The wavy cardboard is arranged to have its corrugations in step at the slit through which the light would be shining. With the long strips in the position shown, the corrugations are exactly out of step at P: this is an analogue of a crest and a trough of the light waves combining to produce no light: hence a dark band. (When two water waves combine in this way—that is, half a wave-length out of phase—a similar effect can be observed.)

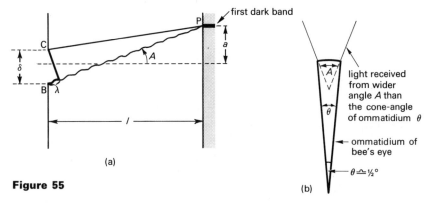

Figure 55

(a)

(b)

first dark band

light received
from wider
angle A than
the cone-angle
of ommatidium θ

ommatidium of
bee's eye

$\theta \simeq \frac{1}{2}°$

The path difference must be a wave-length (λ) for the dark band to happen. Now for the mathematics of the situation. By geometry, since A, the angle of bending is very small, we have:

$$\frac{\lambda}{\delta} = \frac{a}{l}$$

From this: $\lambda = \delta \sin A$, by simple trigonometry. But as A is tiny, $\sin A = A$, very nearly. So $\lambda = \delta A$ or $A = \lambda/\delta$.

Playing God for a moment, we must somehow find a way to strike a nice balance between these two contrary effects. We must find a particular size of eye cell—the distance across δ—to minimize the joint effect. We might do so by drawing a graph or, more elegantly, by using the differential calculus. We add the two effects together and find the value where their sum is smallest. We differentiate thus:

$$\frac{d(\theta + A)}{d\delta} = \frac{1}{r} - \frac{\lambda}{\delta^2} = 0$$

We make the equation equal to zero, to get the 'smallest' sum. School algebra will suffice to rearrange the symbols this way:

$$\delta = \sqrt{\lambda r}$$

How big is δ the size of the ommatidium? To find out, let us guess that the bee's head has a radius (r) of about 4 millimetres. We know from many experiments on colour vision that the bee sees ultra-violet light, which has a typical value of 4000 Ångström (a tiny unit) or 0.4 millionths of a metre. Putting these 'guestimates' in our formula, we get:

$$\delta = \sqrt{(4 \times 10^{-3} \times 4 \times 10^{-7})}$$
$$= 4 \times 10^{-5} = 40 \text{ microns}$$

Naturalists tell us that the bee's eye cell is about 30 microns across, and this, as Feynman declared when he worked this calculation, 'is rather good agreement!'

Now if we put in this optimum value of δ in the formula for the angle A, we find $A =$ about 0.6°. This means that a bee cannot distinguish between light from two sources that are less than about 0.6° apart. So it sees space as clumped together, as we do, though less finely differentiated. Nevertheless, there is a limit to this fineness, which bears on the development of a modern model of vision, and of the brain, which we look at next.

A model of the brain

How can we explain the many varied and extraordinary effects of perception? What sort of model can do this for us? Probably a mathematical model, which must somehow match up two pieces of mathematics—one to describe the memory and the thinking of the mind, and the other to describe the anatomy and electro-chemical working of the brain. The matching up is, in fact, an isomorphism, such a powerful weapon in the mathematician's armoury. We now look briefly at a progress report of a topological model of the brain developed by Zeeman, who has approved the authors' simple exposition which follows.

Zeeman's model is provoked by the simple question: 'Do we see things as a two-dimensional Euclidean drawing with one eye?' The answer must be a firm 'No'. For a start, there are an infinite number of points in a Euclidean plane, while *we* can only see a limited number of points. Moreover, we can only distinguish points that are not too close together. So we 'clump' visual information which on closer inspection shows itself to be much like a newspaper photograph, composed of dots. Details smaller than the screen of dots can never be depicted in any photograph. Thus, Zeeman proposes, we are 'led to the notion of a "tolerance" within which we allow an object to move before we notice any difference'. The same sort of situation arises with double pin-pricks applied to the back of a blindfolded subject's hand. The subject is asked to say how many pin-pricks he can perceive while the pair of pin-pricks are made successively closer and closer. There comes a point at which the pin-pricks are so close as to be indistinguishable and are felt as one jab. Most of us have a pin-prick tolerance of a fraction of a millimetre.

Zeeman advances the notion of a tolerance for visual acuity in the same way. He uses the symbol $\sim$ in the sense that if two spots A and B are indistinguishable to the eye, then $A \sim B$. Drawing on the section *Logical relations* in Chapter 3, we may check that this *tolerance* is both reflexive and symmetrical, but not transitive. Now, if we cannot tell A apart from B, can we tell B from A? $A \sim B$ implies $B \sim A$, and this means the tolerance relation ($\sim$) is symmetrical. And as we certainly cannot distinguish A from itself, or $A \sim A$, then the relation ($\sim$) is reflexive. The relation is not transitive, however; for even if we cannot tell A from B (that is, $A \sim B$) or B from C (that is, $B \sim C$), A and C may be distinguishable, so we cannot write $A \sim C$. In symbols, $A \sim B$ and $B \sim C$ does not mean $A \sim C$.

Any perceptual model must embody two extraordinary features of the eye–brain system—the non-localization of cortical firings for localized (near) images and the 'outline' effect. The first feature depends on the wiring of the eye to the brain (see Figure 56). The wiring explains

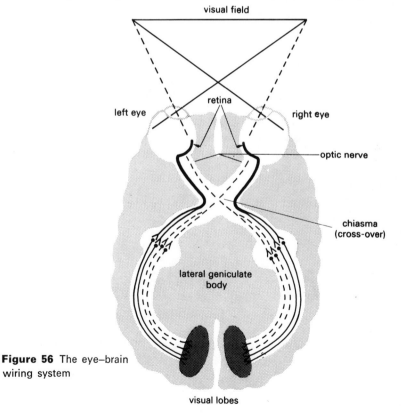

Figure 56 The eye–brain wiring system

why the right half of the brain handles the left half of the visual field. Since the wiring is one-to-one, the visual cortex still represents the same spatial arrangement as the rods and cones in the retina, but naturally much distorted. The rods and cones are packed much more tightly at the centre of the retina than near its edge (see Figure 57). (Our model

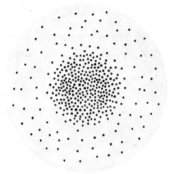

Figure 57 The distribution of optic receptors

takes this into account.) We might think that it would be useful to have the corresponding brain cells in the visual cortex close together where the cells in the retina are close together, and nowhere more so than in the centre of the visual field. But surprisingly, this is just not so. When the eyes see a straight up-and-down line, the right side of the line is seen by the left half of each retina; the optic nerve fibres from the left side of the right eye lead across the optic chiasma where they are met by the optic nerve fibres from the left side of the left eye; they join, like electric cables, and lead to the left visual lobe at the back of the head. Similarly, the left side of the line stimulates the right half of the brain.

Contrary to our intuition which would keep things that are close together in the visual field close together in the visual cortex, we literally have a split-mind over the middle of the visual field. It is almost as if each eye were to take a transparent photograph of the scene it sees, slice the transparency down the middle, flip them open like the flaps of an old-fashioned desk blotter, and register the scene in the cortex in that rearrangement. This wide-split factor can only be handled by a visual model that embodies total non-localization of cortical firings for localized images on the retina.

The second feature of the eye a model must account for is our seeing the outlines of things, a characteristic referred to in the sections on perception in Chapter 2. The eye is equipped with a device to enhance contrast, and thereby the outlines of things. Even children, without any prompting, draw the outlines of their favourite toys and other objects. Of course, there are not lines round things; only our physiological make-up makes us 'see' an outline in this way.

Experiments on the ommatidia of a horseshoe crab have shown that if we shine a pin-point of light on one ommatidium it fires after a pause; but if we simultaneously light another ommatidium close by, then the firing of the second stimulated ommatidium inhibits the firing of the first. This inhibition is greater, the closer together are the ommatidia. When the eye of a horseshoe crab (or of a human being, for that matter) sees, for instance, a cinema screen, one half of which is bright and the other half pitch black, the ommatidia (or rods, in the case of a human) focussed onto the lighted area are inhibited by their nearby neighbours—except at the black–white boundary (see Figure 58).

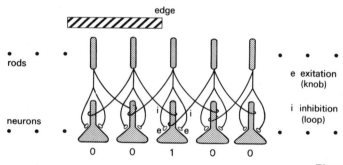

Figure 58

There an ommatidium receiving a light impulse has not so many neighbours firing around it (half are in the dark). Comparatively speaking, this ommatidium sends out *more* impulses along its optic nerve (neurons). For ommatidia on the dark side of the boundary, the presence of others nearby over the border in the light suppress their firing even more, so that the boundary 'looks' even blacker. Surprisingly, scientists have found that in the frog's eye there are some optic fibres that fire continuously even in the dark. From this they infer that there are purpose-built fibres in the eye for detecting specific kinds and changes of light signals.

Zeeman incorporates these features into his model of the brain—actually a model based on topology. But, before looking at this, we will return to the model of visual perception and map the physical situation into a mathematical model (see Figure 59). This model associates with

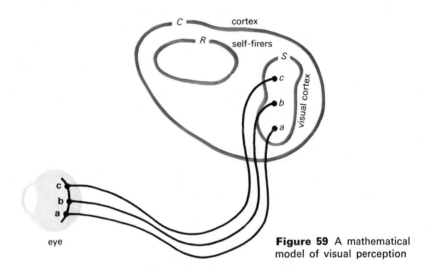

Figure 59 A mathematical model of visual perception

each visual stimulus a 'happening' within the visual cortex. As our picture shows, each retinal receptor is wired to the visual lobe S of the cortex C, depicted as a set on a Venn diagram. The set of visual neurons S is a subset of the set of all brain cells C. A mathematician would write:

$$S \subset C$$

This means that S is wholly included within C. Another subset of the cortical brain cells C are those that fire without any outside stimulus, known as the self-firers, R. They all fire at about ten 'spikes' a second. The name 'spikes' stems from the spiky trace recorded on an electro-encephalogram (EEG); they make up the well-known alpha-rhythm first discovered by Grey Walter. The self-firing cells are like a pilot device that keeps the brain ticking over and alert.

An intriguing interpretation of the Venn diagram follows if we ask what happens if some cells are common to R and S, that is, are both self-firers and are stimulated by the optic nerves. The Venn diagram of

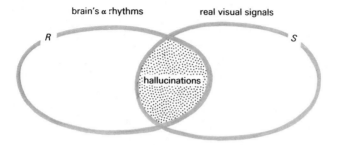

brain's α rhythms   real visual signals

Figure 60

the situation is shown by Figure 60. The overlap of the two Venn rings corresponds to a confusion by the brain of signals sponsored by what is seen and what is imagined, in short, to hallucinations.

If we also consider cells that are fired from within the cortex, and form a three-ring Venn diagram (see Figure 61), the intersection of all three rings on the appropriate Venn diagram might feasibly locate the

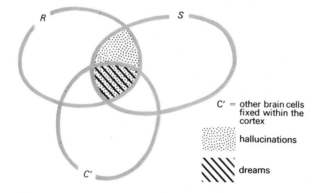

C′ = other brain cells fixed within the cortex

hallucinations

dreams

Figure 61

source of dreams. The currently fashionable theory, that dreaming is like the clearing of its data store by a computer, strikes the authors as too reductionist to be fruitful. A friend, Ensor Holiday, once advanced a beguiling hypothesis about dreams. Dreams at night, he proposed, are evidence of the process of attaching symbols to the meaning of events we have experienced during the day, when we are occupied with asking ourselves 'What does this mean?' His hypothesis suggests a Dormouse-style question to Alice: 'Do you dream what you mean, or do you mean what you dream?' To be sure, the idea of attaching symbols to half-grasped meanings is an attractive starting point for describing the mathematician's flash of insight.

Looking again at Figure 59, let us suppose that light illuminates two receptors in the retina, **a** and **b**, that are within tolerance; for these we may write **a** ∼ **b**. They are both triggered, and both send signals down their own optic nerves to the matching brain cells a and b in the visual cortex. Suppose a is strongly linked by synapses to b, which, in turn, is only weakly linked to another cell, c. By facilitation (a kind of 'knock-on' mechanism), every time the cell a fires it becomes easier for cell b to fire. But a may be quite strongly linked to c. Alone it cannot fire c but, with the help of the 'knock-on' signal via b, c may fire. The model proposes that with education two cells in the cortex S may learn by facilitation to fire in unison. In effect, the cells a and c grow 'closer' together, though by 'close' we do not mean close in the usual Euclidean sense, of course. This suggests that cells a and c may be educated to overlap, to be within tolerance. This is a brief paradigm of learning by association, but it does not explain the deeper processes of abstraction and generalization, discussed in Chapter 2. Thus, before education, we can write $a \sim b$ and $b \sim c$ but not $a \sim c$, because we have agreed that tolerance is not transitive in nature. But after education, we can write $a \sim c$.

We may show facilitation in the kind of diagram de Bono created in his book, *The Mechanism of Mind*. Each line linking the cells a, b, and c represents the strength of the synaptic links (see Figure 62). A strong connection between a and b (strength 5) and a slightly stronger one between a and c (strength 6) and a weak link between b and c (strength 2) still produces the strongest link of all between a and c (strength 6). The links are not chain-like; strengths in series are additive on this model.

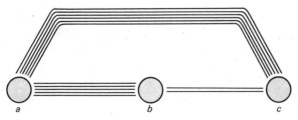

Figure 62

The model also suggests the existence of a reverse map from the cells in the visual cortex to the retinal receptors, which will still trigger receptors within tolerance. Let us think, first, of two points in the mind's eye, that is, on the visual field of one eye, perhaps. The points *a* and *b* correspond to something real the eye has seen. (We had better confine ourselves to an eye of a baby in his first month, for later the human eye develops a finer acuity, called Vernier acuity, after the fine scale on some measuring instruments, which is some thirty times more acute than point—that is, ordinary—acuity.) The mathematician can map two points within tolerance in the mind's eye (set *X* in Figure 63) into an idealization (set *Y*), also mathematical, of the anatomy and chemistry of the brain—a mathematical abstraction. The purpose of this abstraction is to reach a conclusion with the mathematics, by means of the rules of the game, and then to go back to the concrete world of sight

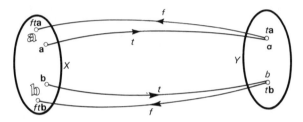

Figure 63

and the brain cells and see if the same conclusion holds. As yet the model has not been put to the test.

In crude mathematical terms we can map from the mind's eye world (*X*) to the brain cells (*Y*) and back again. The 'to' map we denote by *t*, and the 'fro' map by *f*, as in Figure 63. Through *t*, **a** maps into *a* and **b** into *b*. Under the return map, *f*, *a* maps back into **a** and *b* into **b**. As we have said, *a* and *b* are within tolerance, so we write *a* ~ *b*. Since their maps *ta* and *tb* are within tolerance, we write *ta* ~ *tb*. For *ta* is another way, indicating the mapping, of writing *a*, and *tb* is *b* written differently. We can write in our mathematical shorthand that if **a** and **b** are within tolerance, then so are their maps, thus:

If **a** ~ **b**, then *t***a** ~ *t***b**.

Mapping back again, we state that **a** is within tolerance of the original *a*, thus:

a ~ **a** or *ft***a** ~ **a**

Similarly we write for **b**:

b ~ **b** or *ft***b** ~ **b**

What starts out within tolerance ends up within tolerance even after a 'to and fro' mapping. Zeeman develops an entire algebra for handling this kind of situation.

The idea of tolerance extends to language. 'Ideas are sometimes distorted', Zeeman once said, 'by putting them into words, and this process imposes a tolerance on the set of ideas.' Communication between two brains can be mapped like this:

$$\text{ideas} \longrightarrow \text{words} \longrightarrow \text{ideas}$$

The final communication may be inaccurate within the product of the two tolerances. This may be compared to the party game in which a message is passed along a line of people. The more people there are in the line, the more likely the message is to be garbled by the time it gets to the other end of the line. With a really long line, the tolerance may have to be so large as to make the message virtually uninformative. Each language imposes its own tolerance on the speaker: a 'knowledge of two languages', suggests Zeeman, 'imposes a finer tolerance on ideas, thereby sharpening perception'. Let us think of two translations into French of a piece in English, each rendered by a different translator. Each translation may be impeccable in its own right. Yet, one translator's translation must necessarily be inaccurate in terms of the other. The map of the situation looks like this:

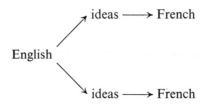

Reassuringly non-reductionist, Zeeman's model rejects the popular notion of the brain as a glorified computer. The brain's shining feature —not shared with the computer!—is its extraordinary ability to store information all over rather than in one place. In this it most resembles the workings of a *hologram*, a wonderful device for taking three-dimensional photographs by special laser light. The mathematician could model the situation as a two-way mapping, like this:

$$\text{scene} \xrightarrow[h]{g} \text{hologram}$$

Information (parts of the original scene) is not stored locally in the hologram. Parts of the holograph, which is like a photographic transparency, may be removed, literally by chopping out a piece of the plastic upon which it is printed, with no other result than to blur the picture. Further, if we select a part of the picture, and map it into the hologram (by mapping *g*) and back again (mapping *h*), we get a 'ghost image' of the total picture. In fact, quite large chunks of the finished holograph can be destroyed without losing the overall picture. And, similarly, the brain can be partially destroyed without impairing thinking.

A dramatic case in point is that of the famous French scientist, Louis Pasteur, who suffered a severe haemorrhage in the right half of his brain early in life. This left him partially paralysed for the rest of his life. An autopsy showed that he had indeed lived most of his life with only half a brain, yet his thinking by any standard was unimpaired. Fortunately Pasteur was right-handed, so that the damage to the right half of the brain did not affect his right hand, which would have been linked with his higher mental activity. This link explains why some left-handed children who are forcibly taught to write with their right hands develop a stutter or defects in their writing or reading. The unnatural use of the right hand stimulates the development of usually dormant centres in the left hemisphere of the brain, thus leading to conflicting signals from the primary centres of both halves of the brain. To accommodate this elusive feature Zeeman invokes the idea of a *tolerance*. However, we should not forget that Rotblat's 'electronic brain' at the Massachusetts Institute of Technology can recover working performance after having its wires cut randomly.

But there is another remarkable aspect of the brain: unlike a pre-programmed computer, it can wander off the point, dream up new ideas, have insights, and behave both directly and vaguely. It is this 'double-think' pattern that has so far eluded computer designers in their 'electronic-brain' making.

Taking as a point of departure, D. O. Hebb's model of the brain, Zeeman associates each thought with a state of the brain. By thinking, neurophysiologists mean actual activity in the neurons and synapses. Sharp thoughts happen when one part of the cortex fires rapidly at one rate, and the other part fires slowly. On the other hand, blurred thoughts arise out of sporadic bursts from many of the cells all over the cortex. Recalling our hologram picture, we are reminded of a laser beam light, produced by electrons jumping orbit within their atoms all together, like a vast symphony orchestra being conducted in strict tempo (or in phase, as physicists would say). That the physicist refers to this 'happening' as coherence may be more than a mere semantic coincidence. Perhaps coherent thoughts are the sharp ones resulting from the brain's orchestra of cells firing in concert and on the beat.

Zeeman pictures a thought as a point on a many-dimensional cube. Naturally, it is impossible to visualize a cube of more than the three dimensions in which we live, although to do so is a favourite trick of physicists who talk about momentum-phase spaces with regard to electrons moving in metals. For simplicity's sake, we shall consider a very simplified version of the Zeeman cube. Where we should represent 10^{10} neurons, we shall restrict ourselves to a mini-model brain with three neurons only. Our cube is a model of all the thoughts that a three-brain-cell man can think (see Figure 64).

The state of each neuron is represented by one of the three axes (edges) of the cube. Each neuron may be quiescent or not firing—state 0 (in the left-hand corner)—or it may be firing—state 1 (another corner). So thought A ('Bell rings', perhaps) could be associated with neuron 1 firing but not with the other two neurons. The thought, then, is shown by the point A on our cube model. If all three neurons fire, the resulting thought ('Dinner time!' say) is shown by the point B on the cube. If the thoughts A and B were to be within tolerance, then the cube brain would associate the bell with dinner. Pavlov educated his dogs to make just such thoughts come within tolerance.

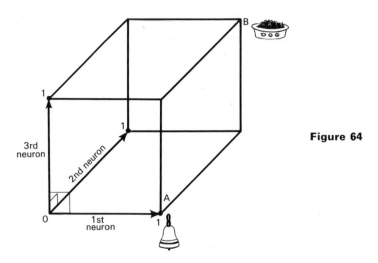

Figure 64

Zeeman pictures a sequential memory, for example, of a favourite tune, as firings moving in a spiral (really a helix because it is not flat) through the brain cube—much like smoke curling up through still air. The direction of thought flow is controlled by the state of all the neurons. We can imagine the thought flowing as along a line on a magnetic field. The elegance of this model is that it allows a thought to start within tolerance, and 'quickly run out of tolerance, implying an element of unpredictability in our thinking and remembering'.

The *sum* of two thoughts—Zeeman calls them t_1 and t_2, the rates of firing of two neurons—may be expressed on his model as:

$$t = t_1 \oplus t_2 = 1 - (1 - t_1)(1 - t_2)$$

Here the sign $\oplus$ means 'putting thoughts together'. Alternatively, he writes:

$$\log(1 - t) = \log(1 - t_1) + \log(1 - t_2)$$

A smattering of modern maths will serve to show that according to this model, thoughts combine associatively and commutatively. An example of associativity in adding ordinary numbers is: $5 + (3 + 2)$, which means $5 + 5$, is the same as $(5 + 3) + 2$, or $8 + 2$. We may pair off the numbers for adding either way. An example of commutativity is: $5 + 2 = 2 + 5$. The way we add commutes, or goes back and forth, as it were.

The model gives a demonstration—stemming from an experiment devised to trick the eye into seeing a fixed scene—of how the finer acuity, the Vernier acuity mentioned earlier, arises. Stare at a point and we find that our eye jerks spontaneously away from the point about every two seconds; in most people the jerk is about seven minutes of arc in angle measure. In the experiment, psychologists fixed a mirror to a contact lens on subjects' eyes, thus correcting the eyes' *saccades* (jerks). After about half a minute subjects found that the scene faded into grey uniformity. This work corresponds with the behaviour of a frog's eye. Some optic fibres fire when it is dark, others fire when it is steadily light, and yet others fire when a light is switched on or off. The signal 'spikes' from the 'steady dark' or 'steady light' fibres or from the 'on–off' fibres are not by themselves strong enough to keep the cortex stimulated. To make up for this deficiency, high-speed boosts (200 spikes a second) from the 'dark' and the 'on–off' fibres lasting about a fifth of a second are needed.

How can the effects of these continual boosts be worked into Zeeman's mathematics? He reasons as follows. An eye-jerk or saccade is one of many rotations the eyeball can make. (Technically, it is actually one of a *group* of rotations.) Over a long period an eye will perform big and little saccades; each size will have its probability of happening. Naturally, gigantic jerks will take place only rarely as will minute ones. Average-sized ones—by definition—will take place more often than not. So we assign to each size of saccade (size s) a chance of its happening, w. Then we sum all these chances up for every possible size of saccade. The sum of this gamut must add up to a dead certainty, to which we give the value 1; while to absolute improbability we assign the value 0.

The sum is in mathematical terms an integration, as follows:

$$\int w \, ds = 1$$

The sophistication lies in how we compute the sum and is signalled for the wary by the letter d, which means, roughly, 'a very small change in'; then ds means a tiny change in s. Zeeman relates a stable thought t— for example, the sight of the pin-point we are staring at—to the boosted thought t^* as a result of a saccadic boost. He links t, t^*, and w in this defining formula:

$$\log (1 - t^*) = \int w \log (1 - t) \, ds$$

Facilitation causes t to grow towards t^*. This means that after the earliest stage in visual education, the thoughts come within tolerance and so are somehow linked.

With this formula Zeeman proves some significant things about perception and mathematics. Straight lines are sharper thoughts than curved lines, as are parallel than non-parallel lines. Outlines are sharper than interiors of shapes (as we now know from neurophysiology), and symmetrical figures than non-symmetrical figures. In the same way, musical chords produce sharper thoughts than discords.

Zeeman's model of the brain, as yet undinted on the anvil of experience, has the outstanding virtue of being provocative and cohesive. At one swoop it embraces such seemingly disparate disciplines as perception and mathematics into a convincing whole.

CHAPTER SIX

Economy

*'But the age of chivalry is gone. That of sophisters, econo-
mists, and calculators has succeeded; and the glory of
Europe is extinguished for ever.'*   EDMUND BURKE

*'Economy is going without something you do want in
case you should, some day, want something you probably
won't want.'*

SIR ANTHONY HOPE HAWKINS:
The Dolly Dialogues

To most of us 'economy' means simply balancing books and budgeting
wage packets, minimizing overheads and maximizing profits, spending
and saving. Managing the household, as we know, seldom requires
more than primary school arithmetic. Until the 1940s, with rare exceptions, business enterprises were conducted by shrewdness, intuition and
elementary accountancy alone. An economist, admittedly, would have
broadened the picture to include such time-honoured generalizations
as the law of supply and demand ('the one creates the other'), 'investment equals savings', and 'unemployment is fractional or voluntary'. If
we add to these the standard textbook topics such as banking, mechanisms of international exchange, the gold standard, business cycles and,
perhaps, public works—Lloyd George's unheeded remedy for unemployment in 1924—the picture is largely complete, with scarcely a
word on unemployment or the Depression!

It fell as we shall see, to John Maynard Keynes, the greatest economist of his age, to change all this. In his masterpiece, *The General
Theory of Employment, Interest and Money*, Keynes rewrote the
language of economics for the Western world. In it he not only de-
molished the bland 'postulates of classical economics' but also replaced
them with a far more realistic (and mathematical) model. Inasmuch as
he was the first man to give serious attention and an answer to such
burning questions as 'What causes unemployment?', he must be con-

sidered a humanitarian of the first rank. For the purposes of the simplified version of his model economy expounded in this chapter, we assume his central point: savings and investment need not be equal. 'It might be supposed,' he writes in his earlier treatise, *The Pure Theory of Money*, 'and has frequently been supposed—that the amount of investment is necessarily equal to the amount of saving. But reflection shows that this is not the case.' Out of this, and other conjectures, he built a mathematical model which, the reader will note, bears a strong resemblance to the latter-day cybernetician's 'feedback' system.

Since Keynes' day, two ideas have radically altered the picture: the theory of games and the digital computer have each stimulated a far more analytical approach to business management. This chapter also outlines the many ways in which mathematics based on these and other ideas are steadily invading the world of commerce and management.

Household expenditure

Household expenditure first received attention over 100 years ago when Ernst Engel published his first paper on Belgian working-class family budgets. His 'Engel curves' served as mathematical models for various domestic expenditure patterns for many years.

Economists classify our household purchases into 'luxuries', 'necessities', and 'inferior goods'. As the income of the normal family increases, it purchases more of the necessities it bought before as well as luxuries it could not previously afford; moreover, expenditure on these luxuries increases faster than income. Figure 65 illustrates this trend with reference to expenditure on holidays. On the other hand, though our expenditure on necessities rises with income, it does so at a slower rate.

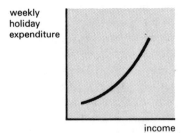

weekly holiday expenditure

income   **Figure 65**

Figure 66(a) illustrates this pattern for expenditure on clothing against income for all four-person families: there we see that the slope of the curve decreases with increasing income.

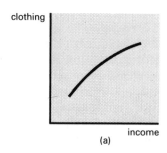

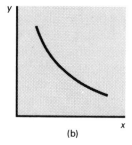

Figure 66

Above a certain income level for, say, four-person households, it may become possible for the family to buy some more expensive substitute for a particular necessity. For example, it may buy butter instead of margarine, steak instead of mince, a car instead of public transport. When this happens the old necessities become 'inferior goods'. At this stage increasing income produces decreasing expenditure on such things (see Figure 66(a)). Engel's model for this was based on the notion that for a given class of households (say, four-person households) the ratio

$$\frac{\text{proportional increase in expenditure } (£y)}{\text{proportional increase in income } (£x)}$$

is a constant (*e*). And thus:

$$e = \frac{\mathrm{d}y/y}{\mathrm{d}x/x} = \frac{x}{y} \cdot \frac{\mathrm{d}y}{\mathrm{d}x} \, *$$

It is interesting to compare typical family budgets in 1900, when expenditure on food accounted for well over half the weekly budget, with those of 1960 when non-food items took most of the money (see Figure 67).

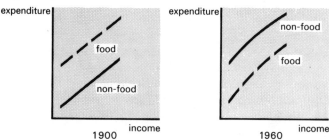

Figure 67

* In Figure 65 the ratio is greater than 1 (*e* > 1). In Figure 66(a) it is positive but less than 1 (0 < *e* < 1). In Figure 66(b) it is negative (*e* < 0). Solution of the differential equation in all three cases gives an equation of the form $y = kx^e$ where *k* is an arbitrary constant.

Investment

Perhaps the family is able or prefers to save, instead of buying luxuries. A great variety of means are available: national savings, building societies, unit trusts, endowment insurance, stocks and shares, and so on. The usual pattern is for some initial principal, £P, to be invested at some rate, $r\%$ (equal to R, when written as a decimal fraction $r/100$), followed by regular (say, annual) payments of £p. What amount of money (£A) accumulates in n years? To start with, in year zero, $A_0 = P$, and in year 1:

$$A_1 = P(1 + R) + p$$

Then in year 2:

$$A_2 = [P(1 + R) + p](1 + R) + p$$

or

$$= P(1 + R)^2 + p(1 + R) + p$$

In year n

$$A_n = \left(P + \frac{p}{R}\right)(1 + R)^n - \frac{p}{R}$$

So, for example, if we put down £100 and pay £100 per annum into a 5% Investment Trust, then after 20 years we should receive:

$$\left(100 + \frac{100}{0.05}\right)(1.05)^{20} - \frac{100}{0.05} \quad \text{pounds sterling}$$

or

$$£[2100\,(1.05)^{20} - 2000]$$

or

$$£3575$$

In the case of house purchase, P is a negative amount, or loan, raised at a little over the bank rate, and £p represents the annual repayment. For instance, suppose we wish to buy a £5000 house over 20 years. The building society may demand a 20% down payment and 6% per annum on the outstanding loan. We could pay in two ways; either we could refund £4000/20, that is, £200 capital each year plus the interest on the outstanding loan, or we could repay some fixed sum each year. By the first method our payments in the early years, perhaps while our children are young and our salary small, would be much larger than those in later years when our commitments might well be much less.

So most people prefer the second method—the way normally adopted by building societies.

What would be the fixed annual repayment £p in the case quoted above?

We have $P = -4000, R = 0.06, A = 0$, and p unknown. Substituting in our formula gives us:

$$\left(\frac{p}{0.06} - 4000\right)(1.06)^{20} = \frac{p}{0.06}$$

or

$$\frac{p}{0.06}(1.06^{20} - 1) = 4000 \times 1.06^{20}$$

or

$$p = \frac{4000 \times 0.06 \times 1.06^{20}}{1.06^{20} - 1}$$
$$= 350$$

So we would repay at a steady rate of about £29 per month.

Depreciation

Suppose a car, purchased new for £1000, depreciates 15% annually. Its value for the next five years will be £1000 $\times$ 0.85, £1000 $\times$ $(0.85)^2$, ... £1000 $\times$ $(0.85)^5$; that is, £850, £722, £614, £522, and £444 successively. Figure 68 illustrates the depreciation graphically. If, however,

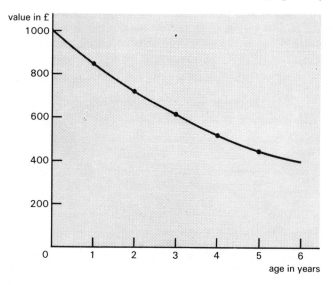

Figure 68

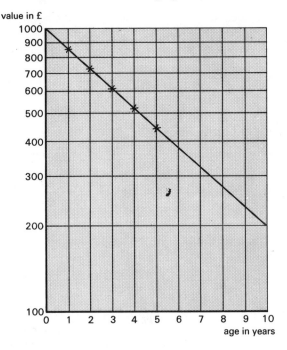

value in £

Figure 69

we plot the values on semilog graph paper as shown in Figure 69, we get a straight line, which is far more useful for interpolating or estimating intermediate values. Of course, any growth law, including the accumulation of money invested at compound interest, appears as a straight line on this type of graph paper.

Nominal and effective rates of interest

Up to now we have been assuming that interest or depreciation are calculated annually. But of course, if a building society decides to compound the interest on our loan quarterly or monthly, life will be far more expensive for us! However, as an investor we would benefit. Suppose that on a sum of money, £p, interest is calculated at a *nominal* rate of $r'\%$, m times per year. With what *effective* rate of interest, $r\%$ per annum, compounded annually, would this amount to the same thing? Obviously, £A, the amount after one year, in either case, is given by:

$$A = p\left(1 + \frac{r}{100}\right) = p\left(1 + \frac{r'}{100m}\right)^m$$

This implies that:

$$r = 100\left[\left(1 + \frac{r'}{100m}\right)^m - 1\right]$$

For small values of r', $r'/100$ will be small, and hence we may approximate by the binomial expansion:

$$r = 100\left[1 + \frac{mr'}{100m} + \frac{m(m-1)}{1.2}\left(\frac{r'}{100m}\right)^2 \cdots\right] - 100$$

Then

$$r \simeq r' + \frac{(m-1)r'^2}{200m}$$

So if $m = 4$ and $r' = 5$, we have:

$$r \simeq 5 + \frac{3 \times 25}{200 \times 4} = 5\tfrac{3}{32}$$

Thus, interest compounded quarterly at 5% per annum is effectively the same as compounding annually at $5\tfrac{3}{32}$%.

Balancing the economy

Attempts to apply mathematics to the whole economy have been made with varying success since the end of the seventeenth century. In 1679 Sir William Petty, the first 'economist' to be elected to the Royal Society, launched his new science, 'Political Arithmetic', with the object of collecting quantitative knowledge on economic, social and political life. Giovanni Ceva wrote the first mathematical book on economics in 1711, but the first really substantial treatise, *Recherches sur les Principes Mathématiques de la Théorie des Richesses*, was published in 1838 by Augustin Cournot. The first Englishman to recognize the importance of Cournot's work and writings was William Stanley Jeavons, whose own *Theory of Political Economy* was published in 1871. Jeavons writes:

'If then, in Economics, we have to deal with quantities and complicated relations of quantities, we must reason mathematically; we do not render the science less mathematical by avoiding the symbols of algebra —we merely refuse to employ, in a very imperfect science, much needing every kind of assistance, that apparatus of appropriate signs which is found indispensable in other sciences.'

Not all economists held this view and, indeed even until recently, many distrusted applications of mathematics on the grounds that economic relationships are generally too complex to symbolize.

It was Hume and Locke who originally put the theory of money on a mathematical footing. Newton, when Master of the Mint, used their theory to stabilize the English currency at a difficult time. He noticed a relation between prices and the amount of money in circulation, subsequently formalized in the so-called 'quantity theory' of money: if the amount of currency in circulation is doubled, other things being equal, then prices—also in terms of currency—will roughly double. In symbols this comes out as the famous equation $MV = PQ$; that is, the total money in circulation (M) multiplied by its rate of circulation (V) equals the total prices (P) times the quantity of receipts (Q). Strictly, PQ should be the sum of lots of individual income transactions involving receipts—the price of apples p_1 times q_1, the number of receipts therefrom, plus the price of houses p_2 times their receipts q_2, and so on —these the mathematician symbolizes as $p_1 q_1 + p_2 q_2 + \ldots$ or as $\sum pq$, which is known as the total 'impulse' to the economy. It is tempting to suggest that Newton was led to this formulation by analogy with what the student of mechanics will recognize as Newton's law connecting impulse and momentum. Provided, then, that the rate at which money circulates (V) and the quantity of receipts (Q) are steady, the formula shows that the money (M) and the prices (P) should rise together. All modern theories of economics presume—and not always justifiably—that a continued rise in M will inevitably expand the economy by giving it an impulse PQ. It won't of course if money circulates less.

The application of mathematical logic to the general principles of the science had to await the genius of John Maynard Keynes and John von Neumann. Keynes, to whom reference has been made at the beginning of this chapter, was a great economist and a highly successful financier; through public office and private writing he wielded a great and lasting influence on fiscal policy. Von Neumann's contribution, the theory of games, though more theoretical, represented a promisingly fruitful and entirely original line of attack.

Simple model of a closed economy

It is worth beginning with the very simplest mathematical model of a closed input–output economy. Suppose we have four countries, A, B, C and D, who trade only among themselves. We will assume also that if any country exports £x worth of goods to another it receives in exchange funds of value £x. Suppose, in our quadripartite common market, that the flow of funds is as indicated by the network shown in Figure 70. In other words, there is no accumulation of funds in any

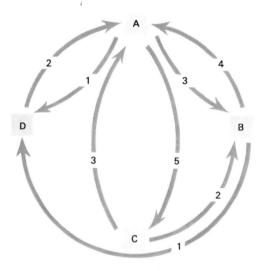

Figure 70 The flow of funds in a quadripartite common market

country—this is rather similar to an electrical circuit, in which, by Kirchhoff's law, there is no accumulation of current at any one junction. Or, we could represent the same situation as an economy matrix:

$$
\text{Flow from} \quad
\begin{array}{c}
 \\
A \\
B \\
C \\
D
\end{array}
\begin{array}{c}
\overset{to\to}{} A \;\; B \;\; C \;\; D \\
\left(
\begin{array}{cccc}
0 & 3 & 5 & 1 \\
4 & 0 & 0 & 1 \\
3 & 2 & 0 & 0 \\
2 & 0 & 0 & 0
\end{array}
\right)
\end{array}
$$

A rectangular or square table or array of figures of this kind occurs very frequently in modern mathematics and is called a matrix. One of our most powerful mathematical tools, it was invented by Arthur Cayley in 1858.

This book can merely indicate some elementary properties and uses of matrices. In our economy matrix, we notice that for any country, say B, the payments to B from A, C and D as shown in the second column, total 5, while the payments from B to A, C and D, as shown in the second row, also total 5. For a closed input–output economy we should, of course, find that the sum of entries in any row (say, the *n*th) was equal to the sum of entries in the *n*th column. In fact, with a larger system of *n* countries, a mathematician would write this rule or condition in symbolic form. For the flow of money, or goods, from country number 1 *to* country number *r*, he would perhaps write f_{1r}, and thus the total payment to country *r* is:

$$f_{1r} + f_{2r} + f_{3r} + \ldots + f_{nr} = \sum_{p=1}^{p=n} f_{pr}$$

Payments *from r* to all the other countries would total:

$$f_{r1} + f_{r2} + f_{r3} + \ldots + f_{rn} = \sum_{q=1}^{q=n} f_{rq}$$

And so for well-balanced books we should have:

$$\sum_{p=1}^{p=n} f_{pr} = \sum_{q=1}^{q=n} f_{rq}$$

In any practical situation, for example, a self-supporting, independent country which needed neither to export nor import, or a closed trading community of countries trading exclusively among themselves, the actual matrix showing the financial incomings and outgoings of every single person, firm and corporation, would be huge and immensely complicated. We can make a very simple, rough 'model' of what is happening by considering three main divisions (or subsets) of all the accounts involved. Let us call them the production accounts (P), consumption accounts (C), and savings accounts (S). These three sets are mutually exclusive (or 'disjoint', as the mathematician would say), but accounts of any given firm feature in more than one category. For instance, a firm which is primarily a producer of motor cars will have an account in P, but as a consumer of accessories or fuel or food, some

Figure 71

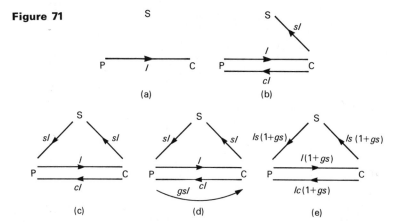

(a) (b) (c) (d) (e)

of its accounts will also feature in C. Very roughly, P, C, and S correspond to Business, Households, and Banks. Let us look at what happens in such a set-up.

Figure 71 illustrates the process. In Figure 71(a) there is a flow of national income I paid to the set of all consumers (that is, everyone in the community) from the proceeds of the sale of consumer goods made by P, here supposed to be the only wealth-producing source. In Figure 71(b) we see that s, a fraction of I (that is to say, sI) is saved in the accounts in S, while the remainder, c, a fraction of I (that is, cI) is paid back to P for the consumer goods received. The money saved in S is now borrowed for capital investment purposes by P. We thus have in Figure 71(c) a closed input–output system. We should notice that consumption (cI) and savings (sI) together exactly equal the national income I. Hence $c + s = I$. Also, we have assumed that all the money saved, sI, is subsequently invested in industry. Now let us suppose that this investment stimulates production, and that the investment capital flowing from S to P results in additional production equal to a proportion, g, of the money invested; that is to say, the national income increases by gsI. The total flow of income from P to C now becomes $I + Igs$, or $I(1 + gs)$ as shown in Figure 71(d). As a consequence, more consumer goods are purchased [$cI(1 + gs)$], more money is saved [$sI(1 + gs)$], and thus even more money becomes available for investment in P. So at Figure 71(e) we find that the system is again closed. If this cycle continues to repeat itself and the fractions c, s, and g remain constant, we have a constant growth rate of gs times the previous production level, or $(100gs)\%$. For example, suppose that $\frac{1}{10}$ of I is saved (that is, $s = \frac{1}{10}$), and that the output yields $\frac{2}{5}$ or 40% of capital invested (that is, $g = \frac{2}{5}$), then the growth rate is $(100 \times \frac{1}{10} \times \frac{2}{5})\% = 4\%$.

It would appear that the growth rate can be increased simply by increasing s or g; that is to say, by saving a larger proportion of income, or by increasing the output-to-capital ratio, or by doing both. If, for instance, either s becomes $\frac{1}{8}$, or g rises to $\frac{1}{2}$, then the growth rate improves to 5%. However, if both rates alter simultaneously, the growth rate becomes $(100 \times \frac{1}{8} \times \frac{1}{2})\% = 6\frac{1}{4}\%$.

The significance to an undeveloped country of hoisting the growth rate from 4% to $6\frac{1}{4}\%$ is clearly shown by the fact that £100 invested for 30 years at 4% amounts to £320, while at $6\frac{1}{4}\%$ it grows to £620—nearly double the first amount.

Of course this little model is too good to be true, as the world found to its cost in the depression of the 1930s. It takes no account of the role of the government, nor of the effect of imports and exports. More important, it suggests a perpetuation of the classical view that savings invested are inevitably used to increase productive capacity and that this in turn tends to increase the national income; in short, that thrift leads to prosperity. There are conditions in which this simple cycle can operate, but these were patently non-existent during the depression. Politicians and economists alike were baffled by the problems of unemployment and over-production.

Escape from an impasse

A way out of the theoretical impasse was shown by John Maynard Keynes (1883–1946), who was conceivably the most versatile economist in England since Sir William Petty. Keynes was the son of a Cambridge don. Educated at Eton (where he made a speech on the thesis that 'women are more fitted to rule than men'!) and King's College, Cambridge, where later he became a fellow, he enjoyed a dazzling academic career in classics and mathematics. It was the noted economist, Alfred Marshall, who decisively persuaded Keynes 'to give up everything for economics'.

Keynes served for two years at the India Office and then wrote as a result of this experience his first major book, *Indian Currency and Finance*. Characteristically, he abruptly resigned as a British Treasury advisor at the Versailles Peace Conference over what he considered to be the Allies' fumbling attempts at peacemaking through reparations. In his 1919 *Economic Consequences of the Peace*, he glimpsed the precariousness of post-Victorian *laissez-faire* arrangements and predicted the depression. It was this book, written at white heat, as an indictment of the Versailles settlement and 'to show that the Carthaginian Peace is not *practically* right or possible', which first brought him fame. It also brought on him 'great odium in official circles' and cast him 'for many years in the wilderness'. His greatest achievements, *The Pure Theory of Money* and *The General Theory*, lay ahead. Chairman of many enterprises, he made a fortune as a speculator. He married the Russian ballet dancer, Lydia Lopokova. Wealthy and immensely successful, he patronized art and the ballet. Eventually he was appointed Director of the Bank of England. His death in 1946 left the country—and perhaps the world—with no economist of comparable stature.

We shall not look in great detail at Keynesianism, which is certainly a controversial issue, but only at some of its simpler tenets. Keynes' central doctrine was twofold. First, savings are determined by investment in productive capacity and not the other way round. Secondly, the national income, I, depends on the 'propensity to save', s, according to the relation:

$$\text{Investment } (S) = sI$$

or

$$I = \frac{\text{investment}}{s}$$

In other words, at a given level of investment, if s suddenly increases (that is, if people suddenly start to save a greater fraction of their income), then I, the national income, will decrease. This is not difficult to understand if we imagine the sequence of events. Suddenly the householder saves more of his income and therefore spends less than before on consumer goods. The producer finds that there is more money available for investment, but that he is selling less than before. He has two choices: he can continue to invest in his factory in such a way as to improve his products and lower their cost, and so win back trade, or he can play safe and decide to discontinue expanding his productive capacity for the time being. Indeed, he may even trim his production to the decreased demand and lay off workers. If he does this, and many would, there is a fall in the wages paid (that is, the national income). Thus, in Keynes' view, greater thrift does not *per se* create prosperity; indeed, it may have the reverse effect. Neither the national income nor the national product are determined by productive capacity; they fluctuate in such a way as to keep savings equal to investments. Before

Keynes the economists had generally encouraged thrift at all times. In his *Treatise on Money* Keynes held that it was 'most desirable in times of incipient inflation but not at all times; on the contrary, in times of depression and unemployment, it was desirable to encourage spending and lavishness'.

We may illustrate these ideas graphically. The national income I consists of two parts: one spent on consumer goods which we shall call C, and which is equal to cI, and the other saved, which we shall call S or sI. Then $I = C + S$. Also, since what is saved, S, is available for investment V, we have $S = V$. Hence:

$$I = C + V \qquad (1)$$

Now we know that $C = cI$, and so:

$$C + V = V + cI \qquad (2)$$

The graph in Figure 72 shows equations (1) and (2) with $C + V$ on the vertical axis and I on the horizontal. Obviously, as $I = C + V$, the graph of (1), $C + V = I$, is the straight line OA at 45° to either axis. The other equation, (2), $C + V = V + cI$, makes an intercept, V, on the vertical axis and has a gradient of c. At the point A, where the two lines cross, the economy is in equilibrium and we have:

$$I = C + V = V + cI$$

or

$$I(1 - c) = V$$

or

$$Is = V$$

So $S = V$ or $I = V/s$ as previously assumed. Now if s, the proportion of income saved, suddenly increases, then obviously c, the gradient of the graph PA, will decrease and the equilibrium point will move to B.

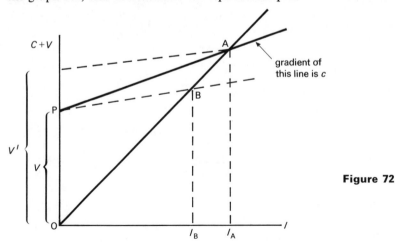

Figure 72

In other words, the annual income will reduce from I_A to I_B. However, the annual income could be retained at its previous level provided V, the savings invested in increased production, were increased to V'; that is, provided producers can be persuaded to invest in increased production capacity at a time when the consumer market is contracting; otherwise, as we have seen, the effect of increasing s is likely to result in a diminished national income.

Still, our model is far too simple; for instance, we have taken no account of exports or imports. Let us now suppose that we import £M of goods and export £X worth. If the proportion of national income spent on imports is m and, as before, the proportion saved is s, then we have: $S = sI$ and $M = mI$. The inflationary components—that is, those that put money into the economy—are now made up of V and the proceeds of exports, X. The deflationary components—that is, those that take money out of the economy—are now the savings S and the imports M. For a healthy balance we require that $V + X = S + M$; that is, $V + X = sI + mI$. So now:

$$I = \frac{V + X}{s + m} \quad \text{instead of} \quad I = \frac{V}{s}$$

This gives us more room for manœuvre.

It is no longe necessary that $V = S$. For suppose that we have that not uncommon situation, an adverse trade balance, with the cost of imports exceeding what we earn through exports. Then $M > X$ and $S < V$; that is to say, investments in production could exceed savings. In the reverse situation, with $S > V$ and $M < X$, we could, for a time at any rate, enjoy a trade surplus. Ideally, $M = X$, so in a deflationary situation it might be necessary to discourage excessive X by raising the rate of exchange. In the last resort each country has considerable interest in the balanced financial development of its neighbours, and this is particularly true in the case of developing countries such as the new African states.

A better model still would have to take account of one more inflationary source—government expenditure G—and one other deflationary drain on the economy—taxes T, levied by the government. If the proportion of national income I removed from the economy by taxation is t, we have $T = tI$. And our new equation of economic balance becomes $V + X + G = S + M + T$; that is, $V + X + G = sI + mI + tI$. So now:

$$I = \frac{V + X + G}{s + m + t}$$

Of all these items, *T* is clearly the factor over which a government has most control—hence the term, *economic regulator*. Moreover, the constant fluctuation of the other quantities offers an explanation of the Chancellor's preoccupation with his annual 'budget day' adjustment of taxation rates. The extraction of *T* from the economy is commonly regarded as a tiresome if not painful process, but it does not, in fact, represent the burden on the community it is sometimes thought to be. For suppose that suddenly all taxes were abolished; higher spendable incomes would immediately begin to pursue an unchanged quantity of consumer goods with a consequent rise in prices. In the end, the burden of *T* would fall most heavily on those least able to afford the increased cost of living.

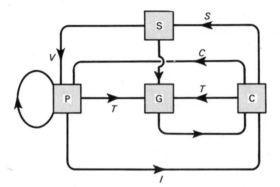

Figure 73

Figure 73 is a flow diagram of the circulation of money between the various accounts: P (production), C (consumption), S (savings), and G (government). Or, reverting to our earlier device, we can equally well represent this network by this matrix, where 1 and 0 indicate whether or not a flow of money takes place in the direction specified.

$$
\begin{array}{c}
\quad\quad\quad\quad\ulcorner to \rightarrow \text{P} \ \ \text{C} \ \ \text{S} \ \ \text{G} \\
\textit{Flow from} \quad
\begin{array}{c}
\text{P} \\ \text{C} \\ \text{S} \\ \text{G}
\end{array}
\left(
\begin{array}{cccc}
1 & 1 & 0 & 1 \\
1 & 0 & 1 & 1 \\
1 & 0 & 0 & 1 \\
0 & 1 & 0 & 0
\end{array}
\right)
\end{array}
$$

Labour control

It is a relatively simple matter to regulate the supply of any section of the professional or labour force by economic means. Increased purchase tax on cars, for example, will reduce home demand and create redundancy among car workers unless all the surplus cars can be exported. In fact, there is a clear relation between the size of a country's trade surplus and the number of its unemployed; sometimes one can actually be calculated from the other.

Let us look at the supply of manpower in one particular profession, education—although the argument applies to almost any job—and consider the factors controlling the supply of teachers in this country. If we suppose N is the number of practising teachers, qN the number of students who have just qualified, t the fraction of those newly qualified students who are to become teachers and r the fraction of all practising teachers who are about to resign, retire or leave the profession in some other way, then we can say that:

(1) the number of newly qualified teachers entering the profession $= qtN$

(2) the number of teachers leaving it $= rN$

So the net gain in the teaching force $= qtN - rN$. But if we suppose that this net gain is also gN, we may write:

$$gN = qtN - rN$$

or simply

$$g = qt - r$$

From the population growth rate, the proportional growth g in the number of teachers required can be calculated. Now g itself depends on q, t, and r. To increase g we must increase q, or increase t, or decrease r, or do more than one of these things. How can this be managed in practice? Obviously r will be decreased by raising the statutory retirement age, and t will clearly increase if there is a rise in teachers' pay. It is not quite so easy to see how to increase q, the number of newly qualified students per existing teacher, but at least it seems reasonable to assume that this 'productivity factor' depends on good qualifications and high morale. Both of these hinge on pay and so it may be assumed that t and q will increase together in some roughly linear fashion for increasing pay. Figure 74 shows the kind of relation that might exist. Some required value of g has, we'll suppose, been calculated, while r cannot be changed. As $g = qt - r$, or $qt = g + r$, where both g and r

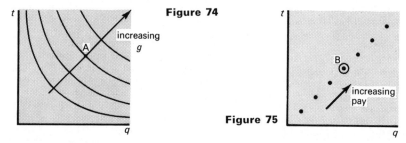

Figure 74

Figure 75

are constant, we see that t and q are inversely proportional to each other for each fixed value of g and r. Figure 75 shows a family of graphs illustrating this—that is, $qt = g + r$ for increasing growth rate g.

Figures 74 and 75 would, of course, be based on statistics available, and they could be used in the following way. We assume that the expansion of the education service requires some particular value of g; r will be known, so we have $g + r$. The appropriate curve now gives us an infinite choice of pairs of values for q and t lying on it. If we fix on the easier of these to control, t, we narrow the solution down to point A, while the corresponding B in Figure 74 tells us the salary levels required to stimulate the approximate growth rate required.

Distribution of wealth and income

In Chapter 8 a certain form of cumulative frequency distribution graph —the Lorenz curve—is used to show the dispersion of population over parts of England. The same technique can be applied to illustrate the dispersion of wealth and incomes (see Figure 76). Lorenz curves show that the distribution of wealth is far more concentrated in Great Britain than it is in the United States: the curve for Great Britain bends farther away from the absolute equality line OP than that for the US.

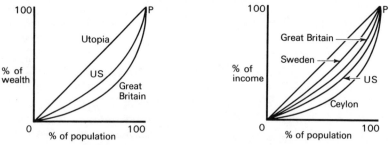

Figure 76 The dispersion of wealth and income

For incomes the pattern is reversed: money from income is slightly less evenly distributed in the US than in Great Britain; distribution is far better in Sweden and far worse in Ceylon. The curves for untaxed and taxed incomes in Great Britain are also interesting. Expectedly, they reveal—as they should!—a far more even distribution of the national income after tax than before.

Saving time

The best way to save time travelling is to fly. Here is another network problem—about air routes. In Figure 77 A, B and C are countries (not drawn to scale!) with airports a_1; b_1, b_2, and b_3; c_1 and c_2. The figures on the straight lines linking the airports represent distances in thousands of miles. So, for example, c_2 is 3000 miles from b_2.

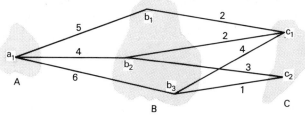

Figure 77

We can again store all this information more compactly in matrix form. Thus, flight routes from country A to country B are given by:

$$\begin{array}{c|ccc} & b_1 & b_2 & b_3 \\ \hline a_1 & 5 & 4 & 6 \end{array}$$

Let us call this F. Flight distances from B to C are:

$$\begin{array}{c|cc} & c_1 & c_2 \\ \hline b_1 & 2 & - \\ b_2 & 2 & 3 \\ b_3 & 4 & 1 \end{array}$$

Let us call this N. The shortest routes from A to C are obviously given by the matrix:

$$S = \begin{array}{c|cc} & c_1 & c_2 \\ \hline a_1 & 6 & 7 \end{array}$$

If $S = F * N$, what is the operation $*$? Well, clearly the shortest route from a_1 to c_1 is the minimum sum of (a_1 to b_1 and b_1 to c_1) or (a_1 to b_2 and b_2 to c_1) or (a_1 to b_3 and b_3 to c_1). Symbolically we may write:

$$S_{a_1c_1} = \min_{m=1\,\text{to}\,3} [f_{a_1b_m} + n_{b_mc_1}]$$

In general, if in a more complicated situation we have l airports in country A, m airports in country B, and n airports in country C, then the shortest route from the lth airport of A to the nth of C is given by:

$$S_{ln} = \min_m [f_{lm} + n_{mn}]$$

Similarly, the longest route is given by:

$$S_{ln} = \max_m [f_{lm} + n_{mn}]$$

The reader may well wonder why anyone should want to know the maximum distance through a network, but if the elements of the matrices were times instead of distances, then the elements of S would give maximum time through the network. In programming large construction jobs it is just this maximum time, giving the *critical path*, which has to be watched, as we shall see in the next section.

Looking again at Figure 77 we may be interested not in the distances but simply in the *number* of different routes between, say, a_1 and c_1 and c_2. Incidence matrices showing these are:

F	b_1	b_2	b_3
a_1	1	1	1

N	c_1	c_2
b_1	1	0
b_2	1	1
b_3	1	1

S	c_1	c_2
a_1	3	2

Notice, that in this case, $S = F.N$ where $F.N$ is the ordinary matrix product of F and N.

Critical path analysis

A youth club has just been offered the long-term lease on a run-down parish hall. One day the club committee decides to do the place up throughout and fixes the grand opening ceremony for twenty-one days later. It even books a teenage star to open the hall. So the work simply must be done on time. Of the many jobs on hand, some must be completed before others can be begun; others can be done at the same time. The final plan of campaign is shown in Figure 78.

At each junction there is a box with two times in it. The left-hand one, obtained by working forward from the start, gives the earliest date on which work can commence, that is, *the earliest event time*. The right-hand figure, obtained by working backwards from the final date, gives the final date on which work may be commenced, that is, *the latest event time*.

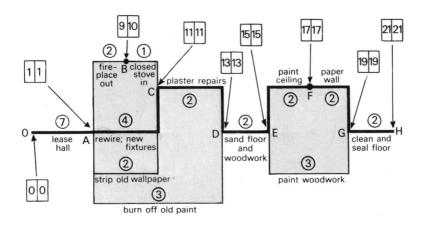

Figure 78 The youth club's decorating campaign

The longest time path through this network is shown as a heavy line. Along this path there is no slack between earliest and latest event times; all activities must start and finish exactly on time if the job is to finish to schedule. This is known as the *critical path* and requires the closest attention. All other paths have 'floats' at their junctions, that is, they have one, two, or three days' grace between the earliest and latest starting times of events along them. The following are the activities with 'floats':

Fireplace out—could start 7th or 8th day after start
Stove in   —could start 9th or 10th day after start
Strip paper   —could start 7th or any day to 9th
Burn paint   —could start 7th or any day to 10th
Paint wood   —could start 15th or 16th day after start

All other activities are on the critical path and must start on time. Of course, due to early delivery, or rapid pace of working, or some loss of time elsewhere in the network, the critical path may change its route through the network; hence the need for running revisions.

The mathematical condition is of the form shown in the previous section, *saving time*; that is:

$$S_{ln} = \max \left[f_{lm} + n_{mn} \right]$$

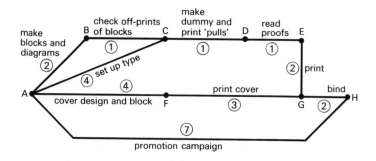

Figure 79 A simple book-production network

The simple network—showing the production of a book—in Figure 79 has been provided for the reader to mark in the earliest and latest event times and the critical path. He might then try to construct examples of his own, and if he has access to the network analysis of some large building project in his own locality, he should study it. This will lead to an appreciation that in a complex situation the critical path only becomes evident after running a computer programme to find it.

Linear programming

In today's highly competitive markets, commercial and industrial enterprises must plan to ensure the highest productivity and profits possible, and the lowest overheads and running costs. The given conditions are often quite straightforward—that is, linear in form. But usually there are so many of them that the calculations have to be done by computer. We now look at a problem illustrating an economic dilemma, which is typical, yet simple enough to be solved by school algebra.

An oil refinery wants to plan its day's production to make the biggest profit possible. It can turn out two products from crude oil: petrol, on which the profit is $2 per Barrel, and jet fuel, on which the profit is $1 per Barrel. And 1 Barrel = 1000 Barrels. These are the operating conditions:

(1) Only 10 Barrels of crude oil are on tap each day.
(2) The refinery must turn out at least 1 Barrel of jet fuel to meet a government contract and deliver it 10 kilometres away.
(3) It must also turn out at least 2 Barrels of petrol for garages and deliver it 30 kilometres away.

(4) Both products are shipped by a fleet of trucks. The fleet can only handle 180 Barrel-kilometres—that is, altogether the entire fleet can convey 180 Barrels over 1 kilometre; or 1 Barrel, 180 kilometres; or 18 Barrels, 10 kilometres; and so on.

What should the refinery produce? We'll suppose it makes j Barrels of jet fuel and p Barrels of petrol. Then we can turn the conditions (1) to (4) into these so-called inequalities (the symbol $\geqslant$ means 'equal to or greater than'; the symbol $\leqslant$ means 'equal to or less than'):

$$\begin{array}{ll} \text{I} & j + p \leqslant 10 \\ \text{II} & j \geqslant 1 \\ \text{III} & p \geqslant 2 \\ \text{IV} & 10j + 30p \leqslant 180 \end{array}$$

We can show these on a graph (see Figure 80).

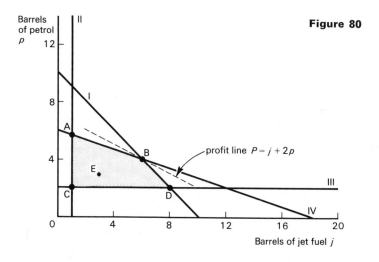

Figure 80

The straight lines are the graphs of I to IV using only the $=$ signs. The shaded area satisfies the equal and inequality signs for all four statements. In other words, the shaded area and its boundary is made up of points that represent all the possible proportions of jet fuel and petrol that the refinery can produce under the given limitations. Point E, for instance, stands for 3000 barrels of each.

But each of these points gives a different profit. For example, the points A, B, C, and D gives us profits as follows:

	j value	p value	Profit = (j + 2p) thousand $
A	1	17/3	12 333
B	6	4	14 000
C	1	2	5 000
D	8	2	12 000

B looks like the most profitable proportion to make. Can we be sure there is not another point which gives an even greater profit? Now the profit (P thousand dollars) is given by:

$$P = j + 2p$$

This is a straight line of gradient $-\frac{1}{2}$, and the further it lies from the origin, the bigger is P. The line for greatest profit is shown as a broken line—and as we can see—gives B as the most profitable programme.

Transport problems

It is possible to plan mathematically bus services for a town, to route services and garage buses at one or other of the bus terminuses. We may picture a very simple town, Squaresville, with a and b as its two bus depots, and A, B, and C the termini of its three bus routes. Each depot can garage three buses. Every morning two buses must be sent to each terminus. Assuming that the buses may travel only North/South or East/West, we want to discover the most economical way of sending out the buses. Figure 81 shows Squaresville.

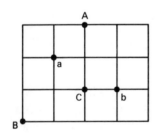

Figure 81

The distances involved are given by the following matrix:

termini

		2 buses	2 buses	2 buses
		A	B	C
3 buses	a	2	3	2
3 buses	b	3	4	1

depots

Now if we suppose that the number of buses to be sent from a to A is x, and from a to B is y, then we have this delivery schedule:

	A	B	C
a	x	y	$3 - (x + y)$
b	$2 - x$	$2 - y$	$(x + y) - 1$

Thus, if M is the total mileage involved in the operation we see that:

$$M = 2x + 3y + 2[3 - (x + y)]$$
$$+ 3(2 - x) + 4(2 - y) + 1[(x + y) - 1]$$
$$= 19 - 2x - 2y \quad \text{or} \quad 19 - 2(x + y)$$

Clearly M is a minimum when $x + y$ is a maximum, and this maximum value of $x + y$ is 3; so the minimum value of M is $19 - 2 \times 3 = 13$. Further, $(x + y) - 1$, the number of buses to be sent from b to C, cannot be negative, so $3 \geqslant x + y \geqslant 1$. The only integral solutions of this consistent with $x + y$ being a maximum are when $x = 1$ and $y = 2$, or $x = 2$ and $y = 1$. Delivery schedules corresponding to these two possibilities are:

	A	B	C
a	1	2	0
b	1	0	2

	A	B	C
a	2	1	0
b	0	1	2

or, in diagrammatic form, as in Figure 82.

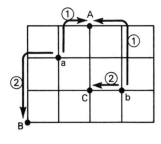 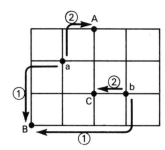

Figure 82

Games theory

In February 1943, General George Churchill Kenney, Commander of Allied Air Forces in the Southwest Pacific, was faced with this problem.* The Japanese were about to reinforce their army in New Guinea and had a choice of two alternative routes. They could sail either north of New Britain where it was rainy or south where it was usually fair. Either way the journey took three days. The expected number of days of bombing exposure was as shown by the matrix:

| | | *Japanese choice* | |
		northern	southern
Allied choice	northern	2 days	2 days
	southern	1 day	3 days

The problem was: should Kenney concentrate his reconnaissance aircraft to the North or South of the island? In the event, both the Allied and Japanese forces chose the rainy northern route. This exposed the Japanese forces to two days bombing. If, however, both had chosen the fair-weather southern route, the Japanese would have been subjected to three days' bombing—and this would have suited General Kenney even better! Similar problems arose in D-day planning in 1944 and with optimum strategy decisions over low and high flying of bombers and fighters.

Such problems in business or war are solved today by one of the youngest branches of mathematics—games theory. It started in 1928 when the Hungarian mathematician John von Neumann developed a rational strategy for the old game of matching pennies. In this, two players, Aristotle Row and Niarchos Column, say, each put down a coin simultaneously. If the coins are alike—both heads or both tails Row takes them; if not, Column wins. The problem is: what strategy should Row adopt?

Later von Neumann teamed up with economist Oskar Morgenstern to produce *The Theory of Games and Economic Behaviour* in 1944. Games theory has as much to do with the players as with the games,

---

* For this story and interpretation the authors are much indebted to Morton Davis's book *Game Theory* (see Bibliography).

and the 'games' themselves include not only 'bluffing', matching pennies, and other simple contests of this kind, but economics and warfare. The theory lends itself to any set-up in which one person seeks a rational strategy, while keeping to a set of rules to best another: the trouble is, the theory cannot actually tell that person how to win, merely the 'best' strategy to adopt! While the main spur to the development of the theory arose from researching its applications to business and war; it would be totally misleading to suggest that games theory is the answer to the prayers of all ardent bridge players or the ultimate deterrent against the grand masters of chess. As we shall see later, the analysis of even the simplest two-person games often involves considerable mathematical complexity. Morgenstern made this point conclusively when he wrote: 'Even a simplified version of poker involving a three-card deck, a one-card, no-draw lead and two participants, would require for its strategic determination the performance of at least two billion multiplications and additions.'

Nowadays the mathematical analysis of strategy in business and war is, at certain levels, becoming quite commonplace. In management courses business situations are simulated and played out by the students as if they are participating in some super game of monopoly. The same kind of activity, applied to warfare, is studied by defence experts. In his *The Bomb and the Computer*, Andrew Wilson describes computer war games in which ace strategists fight tomorrow's battles—and the startling situations they foresee. After one such game, STAGE (Strategic Atomic Global Exchange), 'played' in 1963, the computer produced the 'comforting' (*sic*) answer that in a nuclear war with Russia, the US would 'prevail'! Perhaps the raciest games theory prediction is the one that prophesied India, Africa and China would by 1984 be completely dominated by a new religion, Muluism, based on the tenet that the white man is evil. South Africa would have been rid of whites since a bloody revolution in 1978; south of the Sahara the 'Black African Federation' would be armed with dozens of intermediate-range nuclear missiles, while their ten-million-strong army would still be equipped with spears! Out of a need for mutual security, America, Russia and Europe would have formed a loose economic federation.

But let us turn to the theory itself and start by looking at our two competing commercial giants, Aristotle Row and Niarchos Column, R and C for short. Each has three different strategies open to him, R_1, R_2, R_3 and C_1, C_2, C_3 respectively. Let us suppose further that the profits of R and C, corresponding to the various possible combinations

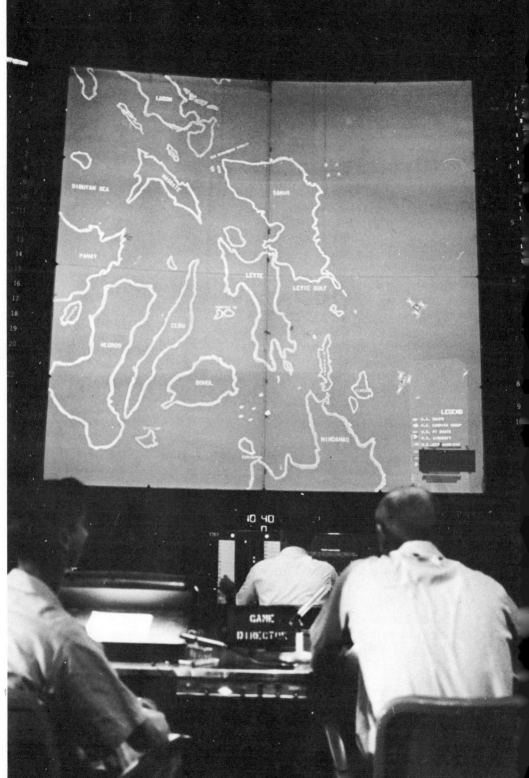

of strategies, may be calculated; these are as shown in the tables or matrices below:

C's possible strategies

$$\begin{array}{ccc} & C_1 \ C_2 \ C_3 \end{array}$$

R's possible $R_1 \begin{pmatrix} 8 & 4 & 2 \\ 6 & 7 & 5 \\ 0 & 3 & 1 \end{pmatrix}$

strategies R_2 R_3

R's profits

C's possible strategies

$$\begin{array}{ccc} & C_1 \ C_2 \ C_3 \end{array}$$

R's possible $R_1 \begin{pmatrix} 9 & 7 & 4 \\ 1 & 2 & 3 \\ 5 & 8 & 6 \end{pmatrix}$

strategies R_2 R_3

C's profits

What will R and C do? Well, being a businessman and not a wild gambler, R is likely to recognize that by adopting R_2 his profit, at worst, is 5, whereas with strategy R_1 he could finish up with a profit of only 2, and worse still, with R_3 he could even fail to make any profit at all. He will, therefore, probably decide to adopt strategy R_2. At the same time C, looking at the possible outcomes of his three strategies, must see that at worse he fares best by adopting C_3. We could say that the solution to this economic 'game' is (R_2C_3), that is, for R to adopt R_2 and for C to adopt C_3. Let us suppose now that C has discovered how R is thinking. C would conclude that R will adopt strategy R_2, and under these circumstances he himself would still be obliged to adopt strategy C_3, so this 'solution' is a very subtle one indeed. In fact, we say that the game is *strictly determined*. The conditions under which this happens may be deduced by careful examination of the corresponding profits 5 and 3 in the tables.

So far we have made a very serious assumption, which may or may not be justified. We have assumed that R and C 'play' against each other and are content to finish with profits of 5 and 3 respectively. But surely it will not be long before both spot that through collusion, that is, by working together, they can do far better. For instance, if they agree that R will adopt strategy R_1 and C strategy C_1, then clearly together they will make a profit of 8 plus 9—or $8\frac{1}{2}$ each! Opportunities for collusion frequently occur both at the roulette table and in the

Figure 83 (*opposite*) War-games training for US Navy officers seen here 'fighting' the battle of the Leyte Gulf originally fought against Japan in 1944. The control team are following and assessing the moves of two teams of young officers kept out of contact with each other in separate rooms. As each team makes a move it is signalled by lights on the panels at the side of the control room, and planes, ships, and submarines are identified as they show up by radar on the screens (*The Observer*)

more solemn atmosphere of company mergers, but mathematically collusion introduces a complicating factor which invalidates our previous thinking. It will not happen, of course, if there is nothing to be gained by it; that is, if the profit sum is constant or zero.

Let us look at a simple two-person, zero-sum game in which two players, Row (R) and Column (C) can each, independently and without the other seeing, pick either $+1$ or -1. After their choices are revealed C pays to R the sum of what they pick. The possible outcomes are shown in this matrix:

$$
\begin{array}{c c|cc}
 & & \multicolumn{2}{c}{\text{C}} \\
 & & 1 & -1 \\
\hline
 & 1 & 2 & 0 \\
\text{R} & & & \\
 & -1 & 0 & -2 \\
\end{array}
$$

By convention all such matrices show the amount paid by C to R, that is, they show R's *winnings*. Now clearly both R and C will realize the possible outcomes. R will think: 'By picking the row containing 0 and 2, *the least I lose* is 0.' At the same time C thinks: 'If I pick the column containing 0 and -2, *the least I lose* is 0.' So R plays $+1$, C plays -1, and each gains 0. We notice that this zero is simultaneously the minimum of the first row and the maximum of the first column. Such an element is known as the *value* of the game.

Now let us suppose that C were to think a little more deeply, as follows: 'I will assume that R thinks as I do. This will lead him to play $+1$. I must then play -1, otherwise he will win 2.' Thinking like this doesn't alter the outcome at all! When this happens we say again that the game is strictly determined, and that R's and C's choices are *optimal strategies*. Since the value of the game is zero, we say that the game is *fair*. This should be obvious; any game in which, by taking thought, one of the participants is always bound to lose, can hardly be fair.

Here is an example of an unfair game. R and C can hold up 1 or 2 fingers; when their choices are revealed, C pays R the total and R pays C the product. The outcome matrix is:

$$
\begin{array}{c c|cc}
 & & \multicolumn{2}{c}{\text{C}} \\
 & & 1 & 2 \\
\hline
 & 1 & 1 & \textcircled{1} \\
\text{R} & & & \\
 & 2 & 1 & 0 \\
\end{array}
$$

R will clearly show 1 finger, and C will probably show 2 (as circled), though he will soon see that it matters little whether he shows 1 or 2. In either case the value of this unfair game is 1.

Non-strictly determined games

Few games, fortunately, are strictly determined. Before looking at one that is not, we should perhaps note that even when an optimum strategy exists and is known to a player who has the intellectual training to apply this knowledge, he still may not make use of this knowledge.

Richard Brayer, an American behavioural scientist, once had several subjects repeatedly play a matrix game with a 3-by-3 pay-off matrix known to the players. The players were told (truthfully) that the experimenter would play randomly. But in fact they invariably played against the experimenter. Not only did they not know how to apply the matrix—and some were mathematics graduates—but more than half actually felt the experimenter was stupid not to use his optimum strategy! On their own admission the players indicated that they had no insight into what was happening. Others have proved, too, that two players playing a game repeatedly but lacking the skill to compute optimum strategies will eventually approach the minimax strategies.

Let us suppose that Row (R) conceals 1p or 2p in his hand and that Column (C) guesses and wins the coin if he is right. The matrix is:

C *guesses*

		1	2
	1	−1	0
R *conceals*			
	2	0	−2

This is neither strictly determined nor fair. It is loaded against R. However, we could dispose of this objection by reversing the roles of R and C every 100 guesses, so that R has a chance to win from C. The interesting question now arises of how R should play and, indeed, how C should guess, for obviously if R argues that by always concealing 1 he will pick the row with minimum loss to himself, then C will soon spot this and always guess 1 and win. Indeed, if R decides on any *pattern* and C spots it, C will play to advantage. But then, even if R realizes this and decides to divide his choices between 1 and 2 pennies in a random manner, is this the optimum strategy, or should he make the choices of 1 and 2 with different frequencies? Here it is probably plain to see that R should pick 1 about twice as often as 2, mixing his choices in a random manner. C, in turn, should realize this and guess randomly 1 twice as often as he guesses 2.

Dominance matrices

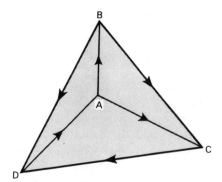

B

A

C

D

Figure 84

Figure 84 illustrates in network form the result of chess matches played among A, B, C, and D. The arrow 'A $\rightarrow$ C' is to be read as 'A beat C'. We can again summarize it in matrix form:

$$
\begin{array}{c}
beat
\end{array}
\qquad
\begin{array}{c}
\text{A B C D}
\end{array}
$$

$$
M =
\begin{array}{c}
\text{A} \\ \text{B} \\ \text{C} \\ \text{D}
\end{array}
\begin{pmatrix}
0 & 1 & 1 & 0 \\
0 & 0 & 1 & 1 \\
0 & 0 & 0 & 1 \\
1 & 0 & 0 & 0
\end{pmatrix}
$$

So A and B have each won 2 matches, and C and D each one. We can differentiate their efforts still further by looking at the cases of *two-stage dominance*. For example, if A $\rightarrow$ B $\rightarrow$ C, then A has two-stage dominance over C. The matrix for this is:

$$
M^2 =
\begin{pmatrix}
0 & 1 & 1 & 0 \\
0 & 0 & 1 & 1 \\
0 & 0 & 0 & 1 \\
1 & 0 & 0 & 0
\end{pmatrix}
\begin{pmatrix}
0 & 1 & 1 & 0 \\
0 & 0 & 1 & 1 \\
0 & 0 & 0 & 1 \\
1 & 0 & 0 & 0
\end{pmatrix}
=
\begin{pmatrix}
0 & 0 & 1 & 2 \\
1 & 0 & 0 & 1 \\
1 & 0 & 0 & 0 \\
0 & 1 & 1 & 0
\end{pmatrix}
$$

This tells us that A has 3 two-stage dominances, while B has only 2. Also, D has 2 to C's 1; hence, the final order is A, B. D, C. Had this not worked we could have differentiated further by evaluating and examining the three-stage dominances given by M^3.

Doing the sums

Many of the processes described in this chapter have been illustrated by very simple examples. Real life is not so simple, instead of 'critical path' networks with fewer than 10 vertices, we find 100 or 1000 or more; instead of linear programming problems involving 4 inequalities, we find 400 or 4000. Doing the real sums is far from simple; techniques of numerical analysis are required and computers are essential.

Throughout history mathematicians have been attacking the drudgery of calculation. Starting with the abacus rooted in the mists of Chinese history (and still in use today) we progressed to Napier's bones and John Napier's epoch-making invention of logarithms and the slide rule in the seventeenth century. Charles Babbage, the nineteenth-century inventor, spent forty years trying to perfect his calculating engine. By 1890 Hermann Hollerith, working in New York, had perfected an electromagnetic, punched card machine—a great step forward. Hollerith machines are still used today, mainly in commerce by hire-purchase companies.

Vacuum tubes were invented in 1900, but the first valve-operated electronic computer was not built until 1946. Since then progress has been phenomenal. Computers are big business all over the world with companies such as the International Business Machines Corporation (IBM), Ferranti, Elliott Automation, and International Computers (ICL).

Pegasus, an early, first-generation, slow, valve-operated computer, could evaluate e from the series $1 + 1 + \dfrac{1}{2} + \dfrac{1}{1.2.3} + \dfrac{1}{1.2.3.4} \cdots$ correct to 1000 decimal places in two seconds. However, in 1948 the transistor was discovered, and the third generation of computers, more compact, more versatile, and much, much faster, can handle fifty such routines in a mere fraction of the time. The last ten years have seen the invention of integrated circuits and their microminiaturization onto silicon chips; in fact present technology permits the incorporation of 100 to 500 integrated circuits into one silicon wafer one inch in diameter and less than .01 inch thick. All this has resulted in a hundredfold cost reduction and an increase in reliability by a factor of 100 to 1000. At the present rate of computer progress every imaginable improvement will have occurred long before A.D. 2000. Indeed, we are already at the limits set by physical factors such as the speed of electric current and the dispersion of heat through plastic materials. Pending break-throughs in solid state physics, present improvements are centred on time-sharing techniques, programme segmentation and parallel pro-cessing computers. Computer languages too are evolving. To read of FORTRAN, ALGOL, MAD, JOVIAL, COBOL, LISP, and CPL might make us suspect that they are also proliferating, but, in fact, the trend is towards a common language, or pair of languages, for most purposes.

Computers are now used widely throughout science and commerce. In *The Year* 2000 Herman Kahn and Anthony Wiener suggest that the following developments are either imminent or clearly foreseeable:

(1) a computerized, national information file storing in easily accessible fashion every citizen's tax, legal and career details, educational qualifi-cations, and medical history;

(2) time sharing of large computers by research centres in every field;

(3) trial configuration testing in scientific work;

(4) a far wider use in business;

(5) instant checking of information, records and fingerprints, which will make possible a vast assault on crime;

(6) computerized banking and instantaneous exchange of money;

(7) medical diagnosis by computer;

(8) traffic control, chemical analysis and weather forecasting by com-puter.

Finally, there is the part computers will play in cybernetics and automated systems. In the United States it is claimed that through automation some 2 to $2\frac{1}{2}$ million jobs per year are being eliminated; in the steel industry alone 80 000 out of the 600 000 have already vanished. This situation does not necessarily imply unemployment, for we have taken no account of the entirely new occupations which will be com-monplace in fifty years' time.

CHAPTER SEVEN

The Mathematics of Maps

'What's the good of Mercatoi   North Poles and Equators,
  Tropics, Zones, and Meridian Lines?'
So the Bellman would cry: and the crew would reply,
  'They are merely conventional signs!

'Other maps are such shapes, with their islands and capes
  But we've got our brave Captain to thank'
(So the crew would protest) 'that he's bought us *the best—*
  A perfect and absolute blank!'

LEWIS CARROLL

'The satellite age for geographers means just one thing—
there must be maths . . . Geographic information will be
coming from lasers, radar beams, even satellites—and we
shall need to process it . . .'

PROFESSOR ALICE GARNETT,
President of the Geographical Association

But first, what is geography and what information is it concerned with? In his *Geography of World Affairs*, John Cole writes: 'Geography is about the earth's surface, and naturally, since man during the last millennia has played a great part in shaping this, the geographer appreciates the close relationship between the physical environment and human activities.' In Chapters 7 and 8 we look at some of the mathematical tools which have been and are being forged in the course of this appreciation. There is no need to elaborate on the amount of mathematics discovered by the star-gazing Babylonians, the land-measuring Egyptians or the contemplative Greeks. Most of what we teach in our school mathematics lessons today was originally motivated by man's need to understand, measure and organize the world around him. A critical point in this development occurred in the fifteenth century. The strangulation of traditional trade routes by the Ottoman Empire provoked a reaction from such emerging and self-conscious states as England, Portugal, and Spain. Through the enterprise of mariners, forced to voyage farther from home, new sea routes were opened up, and this culminated in 1522 with Magellan's successful circumnavigation of the globe. Though it was widely believed in the ancient world that the Earth was a sphere, and though the fact was proclaimed in the sixth century B.C. by Thales, in the fifth century by Pythagoras, and in the fourth century by Plato and Aristotle, its actual physical limits had never before been practically demonstrated. The first globe was not produced until A.D. 1492. Later, the invention of the printing press and the development of the magnetic compass and of firearms provided new means of communication and conquest, by which European thought and influence were conveyed to the corners of the world.

Early maps

Anaximander is often quoted as the inventor of maps, though Democritus, who was born about 450 B.C., approximately a hundred years after Anaximander's death, is believed to have been the first to produce a rectangular map. Indeed, Democritus announced that the Earth was one and a half times as long from East to West as it was broad in the direction North to South. It is from the Latin 'latus' and 'longus', 'broad' and 'long', that our words 'latitude' and 'longitude' derive.

Aristotle quoted the Earth's circumference as 400 000 stadia, a unit of length whose modern equivalent is unknown, so we cannot know how good or bad this estimate was. However, we do know that in the third century B.C., more by luck than good management, Eratosthenes produced a remarkably accurate estimate of the Earth's radius. Lucky or no, his method was mathematically sound and extremely elegant in its simplicity. On Midsummer day the noon Sun, directly overhead, was mirrored in a deep well at Syene in Egypt, now the site of the modern Aswan Dam. On the same day at Alexandria, 500 miles to the north, the shadow of an obelisk was at $\frac{1}{50}$ part of a circle, or $7° 12'$, to the vertical (see Figure 85). Now the Sun's rays are parallel, so the angle at the centre of the Earth between the radii from Alexandria and

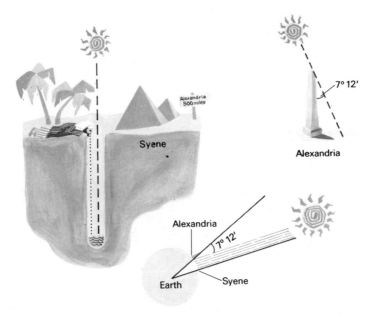

Figure 85 Eratosthenes' estimate of the Earth's radius

Syene is 7° 12'. Taking the distance between the two towns as 500 miles, we have:

$$\frac{500}{\text{circumference of Earth}} = \frac{\frac{1}{50} \text{ of a turn}}{1 \text{ turn}},$$
$$\Rightarrow \text{circumference of Earth} = 500 \times 50$$
$$= 25\,000 \text{ miles}$$

Eratosthenes' result was, astonishingly, within 200 miles of the true polar circumference, 24 860 miles. The fact that he assumed the Earth to be a perfect sphere, overestimated the distance from Alexandria to Syene, and had no accurate means of measuring angles, does not detract from his achievement. Another attempt, by Posidonius two centuries later, using the declination of Canopus from Rhodes and Alexandria, produced a comparable result.

Claudius Ptolemy, probably the greatest geographer of antiquity, who flourished in the second century A.D., left details from which was constructed the world map that remained a standard work until the

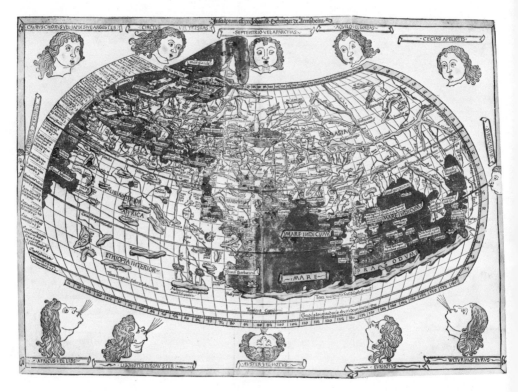

Figure 86 The World from Ptolemy's *Geographia* (*The British Museum*)

days of Columbus. Unfortunately he based his work on an Equatorial circumference of 180 000 stadia—a gross underestimate of the true value.

The great Graeco-Egyptian culture that was centred on Alexandria finally came to an end, in the centuries following the conquest of Egypt by the Moslems in A.D. 640. The Christian world was to enter 'a dark age' during which little scientific progress was made for some six centuries. Meanwhile, with the translation into Arabic of the works of Aristotle and others, Islamic learning continued to develop. Indeed, in A.D. 814, the Caliph Abdullah al Mamun determined the length of a two-degree arc of a great circle on the plains of Mesopotamia.

The movement of the Earth, however, remained a puzzle. Though one of Pythagoras' followers had suggested in the fifth century B.C. that the Earth rotated about its diameter, and Pythagoras himself had suggested that the Earth rotated about the Sun, it was the reputation of Aristotle and Ptolemy and the tradition begun by them, so illuminating in other respects, which really stultified the development of science for so long;

some would say from 250 B.C., to A.D. 1550—1800 years! Aristotle believed the Earth was fixed at the centre of a series of concentric spheres. Ptolemy's view was similar and explained the motions of the Moon and planets by means of cycles and epicycles (sets of circles of different sizes moving one on another). Tailored to fit the facts, the theory was safe from observational disproof. Such was the strength of ecclesiastical support for the views of Aristotle and Ptolemy that all deviations from their Earth-centred philosophy were regarded and treated as heretical. These notions successfully impeded the progress of scientific thought until Copernicus and Galileo challenged them in the sixteenth century.

Such maps as were produced during the Dark Ages were constructed on the basis of religious rather than scientific criteria. The 'T & O' maps (*orbis terrarum*), typical of the times, were circular and showed the world divided into the three known continents, Europe, Africa and Asia, with Jerusalem at the centre. Such a map is shown in Figure 87.

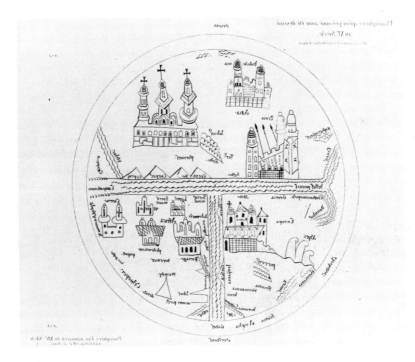

Figure 87 A fourteenth-century 'T and O' map (*The British Museum*)

Figure 88 A section of the Gough map showing the south-east of England (*Ordnance Survey*)

No other maps of note are known before the fifteenth century, except for a Roman 'world map' and one of Britain. A copy of the Roman map, made in A.D. 1230, discovered three centuries later and known as the Peutinger Table, shows Britain to the West and the plain of the Ganges to the East. The most significant early map of Britain, the Gough map, dates from A.D. 1335; today it is kept in the Bodleian Museum. Probably as a result of ecclesiastical influence, East is at the top. Unfortunately its history and authorship are unknown, for it is a remarkably accurate representation for the times as the section in Figure 88 shows.

The dawn of modern, scientific map-making, however, began to break in the fifteenth century, which brought the birth of Gerhard Mercator in 1512 and the development of copperplate engraving. Indeed, Reisch's *Margarita Philosophica Nova* of 1512 featured Martin Waldseemüller's 'polymetrum', which was a combination of theodolite and plane-table. In England the theodolite was also described in the writings of Leonard Digges, while in 1537 the first treatise on surveying, *The Boke of Measuring of Lande*, by Sir Richarde de Benese, was printed.

Saxton of Dunningley

The first English map-maker of note was undoubtedly Christopher Saxton of Dunningley near Leeds. Curiously enough, the registers of the parish church in Leeds record two Christopher Saxtons, one of Bramley and the other of Farnley, but neither was the cartographer. Our nearest estimate of the dates of his lifetime is 1542 (or 1544) to 1608 (or 1610), and it is thought that he was born at Tingley. Educated at Cambridge, Saxton came to London, where he was fortunate enough to win the patronage of Thomas Seckford, Queen Elizabeth I's Master of the Court of Requests and Surveyor of the Court of Wards and Liveries. Seckford recognized Saxton's talent as a surveyor; he encouraged him to survey the counties of England and Wales and paid all his expenses. Saxton's great survey of the English and Welsh counties took nine years. It began in 1570 when he was twenty-six or twenty-eight years old. In 1574 Norfolk, Oxford, Berkshire and Buckingham-

Figure 89 Saxton's map of Norfolk 1574 (*The British Museum*)

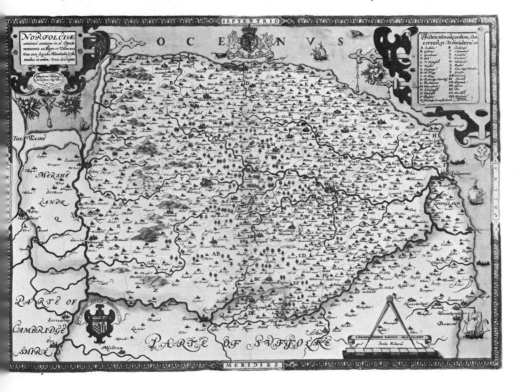

shire appeared as single sheets; maps of the remaining counties were published by 1579.

Saxton was fortunate to have a patron, but to have a patron who could invoke the sovereign's aid and authority undoubtedly accounts for the success and long influence of the survey. In the register of proceedings of the Privy Council of 11 March 1575, Appendix II (III), there is an open letter to the JPs of England, which reads:

'. . . that the said Justices shall be aiding and assisting unto him to see him conducted unto any towre, castle, high place or hill to view that countrey and that he be accompanied with ij or iij honest men such as do best know the countrey for the better accomplishment of that service and that at his departure from any towne or place that he hath taken the view of the saide town do set forth a horseman that can speke both welche and englishe to safe conduct him to the next market towne etc.'

On 22 July 1577, the process of survey, delineation and publication being nearly completed, Saxton obtained from the Queen a ten-year licence for the exclusive publication of his maps.

In 1579, Saxton published the complete collection of thirty-four county maps in atlas form—an event memorable in the history of cartography. The title page, with a portrait of Queen Elizabeth in the centre, was followed by a coloured alphabetical index of the maps with a list of judges' circuits, eighty-three coats of arms (plus one blank shield) of the nobility, a catalogue of cities, bishoprics, market towns, castles, parishes, churches, rivers, bridges, chases, forests and parks for England and Wales, together with a general map of 'Anglia'. This atlas, now very rare, was published at 15s., though later the price rose to 20s. In 1579 Saxton was granted armorial bearings of his own.

In 1584 came Saxton's greatest achievement, a large map of England and Wales which dominated English cartography for 180 years; no original copies remain. Philip Lea cut the plates about and republished the map in 1687, but even of this there are only four known copies extant—at the British Museum, the Bodleian Museum, the Royal Geographical Society and in a private collection. Seckford's death in 1587/8 left Saxton without public employment or patronage, but he was kept busy until his death with private commissions to survey estates.

Saxton's work was almost entirely carried out with a trundle wheel, a plane-table, and a planimetrum, with which he measured the angles of elevation and depression from the top of church towers and hills.

He surveyed in acres (1 acre = 4 roodes), roodes (1 roode = 10 day-works), dayworks (1 daywork = 4 perches), and perches. His survey of the Lordship of Burley, made in 1602, includes the following items (the reader may care to check whether the calculation is correct):

Richard Crosley house and croft	0	1	5	0
The parrock meadow	1	1	0	3
The eastfield close	2	0	6	3
The garth	0	2	4	0
The hay close	6	0	3	0
	10	1	9	2

Other early map-makers

The importance of patronage is well illustrated by John Norden (1548–1625), who had none. Surveyor, inventor of the triangular distance table in his *England*, an intended 'Guyde for English Tra-vailers', and writer of devotional books, Norden's work is generally regarded as being of outstanding quality, and it was far more extensive than that of Saxton. He produced the *Honor of Windsor* for James 1 in 1607, seventeen water-colour-on-vellum plans of great beauty and interest, together with plans of various manors and estates on the Suffolk coast, and in Cornwall and Somerset. Between 1593 and 1598 he also produced and published at his own expense improved county maps of Middlesex, Surrey, and Sussex. His map of Cornwall was not published until 1728, while that of Essex had to wait until 1840. His maps—the first, with Philip Symonson's map of Kent, to show roads—are extremely scarce, but they were used in the *Camden Britannia* of 1607 and by subsequent map-makers such as John Speed. As melancholy and unsuccessful as Saxton had been fortunate and famous, Norden died a disappointed man.

Meanwhile, on the continent of Europe, in Holland and France particularly, cartographical skills were developing rapidly. Two great events of this period were the publication in 1570 of Abraham Ortelius' *Theatrum Orbis Terrarum*, and in 1595 of Gerhard Mercator's complete *Atlas*. Mercator is remembered not only for the term 'atlas', used from that time onwards for a collection of maps, but also for one of the most famous early map projections.

Cartographical techniques at this stage derived mainly from navigational instruments. The astrolabe, about which Chaucer wrote a treatise for his son, the quadrant, the cross-staff and John Davis's improvement of it (the backstaff) were all in use in ships. In the sixteenth century the land surveyor's instruments were nearly all adaptations of these, long familiar to astronomers and navigators. In the first decade the geometrical square was refined into the polimetrum—a combination of compass, quadrant and astrolabe. In 1571 Leonard Digges published his *Pantometria*, an English treatise on contemporary surveying instruments, which included details of the plane-table.

The measurement of distance led to successive improvements of the trundle wheel, and eventually to click wheels and mileometers which could be attached to carriage wheels so that the job could be done in comfort. Even John Ogilby, who in 1670 published the first road atlas, *Britannia*, relied on a measuring wheel just over thirty inches in diameter, which he called a 'perambulator'.

However, by this time more sophisticated devices were available. In 1669 Abbé Jean Picard, who used a chain of triangles to determine the length of an arc of meridian, had a vernier (invented by Paul Vernier in 1631) fitted to the angle scale of his theodolite, and a micrometer (invented by William Gascoine) fitted to its telescope crosswires.

Figure 90 Surveying techniques of the eighteenth century (*Radio Times Hulton Picture Library*)

As a result of using these more sensitive instruments it gradually became evident that the length of a degree of meridian increases as we travel North or South from the Equator. Present-day figures for a degree of longitude at the Equator and the poles are respectively 6046 and 6108 feet. There was only one explanation: the curvature of the Earth's surface is less at the poles than at the Equator; that is, it is flattened at the poles (see Figure 91). Today we describe such a shape as an

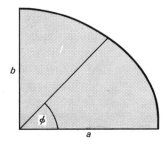

Figure 91

'oblate spheroid'. In fact it is possible to calculate the length of one degree of meridian in any latitude. If we define the ellipticity of the Earth as $c = (a - b)/b$, it can be shown that in latitude ϕ, the length of a nautical mile, that is, one minute of one degree of meridian is:

$$\frac{a}{3438}\left[1 - \frac{c}{2}(1 + 3\cos 2\phi)\right]$$

Thus, in latitude 0°, we have:

$$\frac{a}{3438}(1 - c)$$

where c is approximately $\frac{1}{300}$.

The problem of longitude

When we wish to pin point the position of a place on the Earth's surface—say, Gander—we give two angles, the latitude and the longitude, like this: (49° N, 55° W). The lines of latitude are circles of diminishing radius parallel to the Equator and the lines of longitude are all great circles passing through both poles. Thus, Gander (G) is on a line of latitude where $\angle$GOB = 49° and on a meridian of longitude

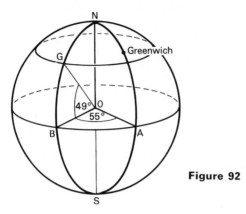

Figure 92

such that $\angle AOB = 55°$ (see Figure 92). Angles of latitude run from 90° S to 90° N; lines of longitude from 0° to 180° East or West of Greenwich.

Of course this convention has not always existed. Latitude was a familiar notion from early times, but agreement over the location of zero longitude and its accurate determination were not achieved until much later. Chaucer, who was born in London about 1340, wrote a treatise on the astrolabe for his son, and in this he described the use of the instrument for determining 'the latitude of Oxford' and other places. Geographers and navigators had for some time been familiar with methods of finding latitude. The most straightforward way was simply to read the elevation of the Pole-star (Polaris).

Figure 93(a) shows the observer at P sighting Polaris. His sight line PN is parallel to ON because the distance of this star from Earth can be reckoned at infinity. Hence the elevation between his horizon and the star, $\angle NPH$, must be equal to the angle of the latitude of P, $\angle POE$.

Figure 93 Finding latitude using the elevation of the Pole star

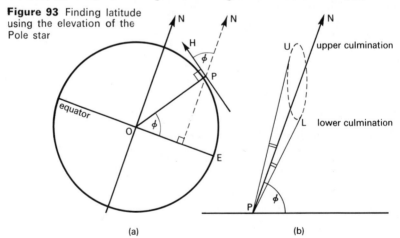

(a) (b)

Alternatively the observer at **P** may select a bright circumpolar star and record its highest and lowest transits or culminations (see Figure 93b). The elevation of the bisector NP of ∠ UPL, corrected for refraction, is the latitude of P. In the southern hemisphere other methods are employed.

The determination of longitude, however, depends on the accurate measurement of time. The way in which longitude varies with time is shown roughly in Figure 94. Since the Earth rotates through its 360° of longitude once every 24 hours, we can see that the local mean time (L.M.T.) at, say, Gander should be ($\frac{55}{360} \times 24$) hours, that is, $3\frac{2}{3}$ hours before the time at Greenwich. So at Gander it should be 2.20 p.m. when it is 6 p.m. at Greenwich. Let us imagine the reverse problem: if we know the local time at Gander, how do we calculate its longitude? This would clearly involve knowing the exact difference in local times between Gander and Greenwich. Given a wireless time signal it is only

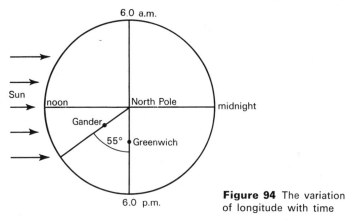

Figure 94 The variation of longitude with time

necessary to have an accurate time recorder at Gander, but before 1759 the seafarer who needed to calculate longitude required two chronometers, one showing local time accurately, the other, which may have been at sea with him for months, showing Greenwich Mean Time. This second chronometer was the problem; it always lost or gained so much as to invalidate all the calculations. There was no accurate time-keeping device available until John Harrison produced his 'mark IV' model in 1759.

For many centuries the problem of the initial meridian remained unresolved. The first serious attempt to clear the confusion came in the 1493 Papal Bull of Demarcation when Pope Alexander VI tried to settle the dispute between Spain and Portugal. He drew a meridian line from pole to pole on a chart of the 'Western Ocean' 100 leagues from the Azores. To Spain he assigned all lands not belonging to any Christian prince West of the line, and to Portugal all discoveries East. Unfortunately, none could determine precisely the position of the line on the ground. In 1612 Galileo devised a method of finding longitude using the satellites of Jupiter, a method which was ingenious but seldom practical, particularly for the mariner. In France in 1634 Louis XIII's government under Richelieu fixed upon Ferro, the Ile de Fer, the most westerly of the Canary Islands, as the point through which the initial meridian should run. This was retained right up to the Revolution in 1789. In England, Saxton based his map work on a prime meridian running through St Mary, the most easterly island of the Azores, and a vague westerly initial meridian was used by most map-makers until 1676 when John Seller, in his map of Hertfordshire, used a zero meridian through London. First St Paul's Cathedral and later Greenwich Observatory were selected as the precise point of origin. There is a good contemporary description of the problem in Herman Moll's *Atlas Monuche* published in 1709. Finally, at an international conference in New York in 1889, it was agreed that the prime meridian should pass through the transit instrument at Greenwich.

Meanwhile two other problems remained: one was the measurement of length, and the other of time. Length, of course, could be measured accurately, particularly after the invention of the micrometer and the vernier scale, but, for example, at the time of Robert Morden's issue of English county maps in 1695, there were in use no less than three different English 'miles' and Morden's maps show scales for all three.

Time, however, was the great problem. One degree of longitude on the Equator subtends 68 miles (109 km), and it is equivalent to a time difference of $1/360 \times 24 \times 60$, or 4 minutes. Thus, 1 minute of time difference is equivalent to 17 miles or roughly 27 km on the Earth. Hence, on a six-week voyage, if the navigator must remain accurate in his plots to $\frac{1}{2}°$ or 34 miles (55 km) his clock must not lose or gain more than 2 minutes in 42 days—or, approximately, 3 seconds a day! No such clock was available, though in 1656 the Dutch mathematician, Christian Huyghens, made an isochronous clock, accurate on land but unreliable at sea. So in 1714 the newly constituted Board of Longitude offered a reward of £20 000 for the first accurate clock invented. A year later the French *Académie* also offered a similar prize. Such inducements attracted many would-be winners and spurred astronomers and instrument-makers to frenzied activity.

Figure 95 Harrison's fourth
chronometer, 1759
(*The Science Museum*)

However, the story of longitude and time-keeping really belongs to the man who finally won the prize. John Harrison moved to Barrow-on-Humber, Lincolnshire, when he was seven years old. Inventor of the 'gridiron' pendulum and the 'non-oil', 'grasshopper' escapement mechanism in the clock, Harrison visited London in 1728 where he saw Edmund Halley (of comet fame) and George Graham, from whom he borrowed money to complete a marine chronometer. The first clock, an unwieldy affair, was finished in 1735, but it possessed the accuracy required and was so certified by Halley, Smith, Bradley, Machin and Graham. In 1736 Harrison applied to the Board of Longitude and the chronometer underwent a successful sea trial to Lisbon in HMS *Centurion*, as a result of which the inventor was voted £500 of the reward. The following year England and Spain were at war, so the improved version of the chronometer, mark II, was never tested. In 1741 work on a third clock started and Harrison was voted a further £500. However, this model developed trouble with the balances. In 1746 Harrison was awarded the Copley Medal. In 1757 two years' work on the mark IV was started. When tested over four months between Portsmouth and Jamaica in HMS *Deptford*, the chronometer lost five seconds over the whole journey! By an Act of Parliament during Queen Anne's reign, John Harrison and his son were nominated winners of the £20 000 reward, but still it was not paid. After further trials, interminable quibbling, the production of a fifth clock in 1770, and finally the personal intervention of King George III, who tested the clock at his own private observatory at Kew when it lost only 4½ seconds in ten weeks, John Harrison and his son were eventually awarded, by Act of Parliament, the balance of the prize money in 1776, a year before the 'Barrow clock-maker's' death at the age of eighty-three.

Calculating distances from latitude and longitude

Given the latitude and longitude of two places (say, A and B), it is possible to calculate the length of the shortest path—or 'geodesic'—between them. This will lie along the line of least curvature, that is, on the circle of greatest radius which passes through A and B. Like the Equator, such a circle will, of course, be a great circle. Assuming the latitude of A as *a* North and the longitude as zero, and the latitude of B as *b* North and the longitude as *P* East, we can prove that the shortest arc AB subtends an angle *d* at the centre of the Earth, where

$$\cos d = \sin a \sin b + \cos a \cos b \cos P$$

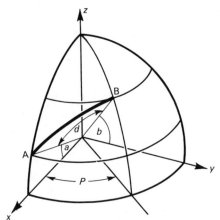

Figure 96

This is shown in Figure 96 where we have a section of a sphere of unit radius. To calculate the arc AB, we simply find angle *d* in radians (a radian is that angle which is subtened at the centre of a circle by an arc of its circumference equal to its radius) and multiply by the radius of the Earth.

The trouble with the geodesic is that it crosses successive meridians of longitude at either increasing or decreasing angles, so to steer along a great circle (except the Equator or a meridian of longitude) involves a constant change of direction. Hence, for example, an aeroplane could not be set to follow a geodesic on automatic pilot. In practice the mariner or navigator prefers to set a constant course. The 'constant direction' line from A to B along the surface of the Earth is called a loxodrome or 'rhumb-line', and such lines have complicated equations.

Figure 97 *(opposite)* Ramsden's theodolite, 1787 (*The Science Museum*)

Birth of the Ordnance Survey

Excellence in one field is often matched by similar development in another. Major-General William Roy, Surveyor of Coasts and Director of Military Surveys in Great Britain, was responsible for the ambitious project, sponsored by the Royal Society, to co-operate with the Cassinis, who were successively in charge of the Paris observatory, in linking (by triangulation) the meridians of Greenwich and Paris. His instrument-maker, Jesse Ramsden, perfected in 1787 a massive theodolite, equipped with telescopic sights and capable of exceptional accuracy. Roy declared it 'extremely perfect'. Instruments of this accuracy and surveyors of Roy's calibre revealed considerable errors in English topographical maps. Maps of greatly improved accuracy were produced, by John Cary and others from 1794 onwards, but it was clearly time for a national survey.

The Ordnance Survey was established in 1791. Initially its maps were of militarily strategic areas and were constructed mainly by army engineers, but later private map-makers, such as Cary, acted as consultants to the Survey. Cary was, in fact, engaged by the Government in 1794 to survey 10 000 miles of roadway. The first one-inch, uncoloured, engraved map produced by the Survey was published on 1 January 1801. Ordnance Survey maps were produced at intervals throughout the nineteenth and twentieth centuries. They were based upon a careful triangulation of the country and a linked triangulation with France. In England eight base-lines were used at one time or another; two of these were on Romney Marsh and Hounslow Heath. The base-line at Hounslow Heath was measured in 1784 as 27 404.01 feet, in 1791 as 27 404.24 feet by Ramsden, and in 1858 as 27 406.19 feet. The two principal base-lines for the Survey were, however, laid down at Lough Foyle in 1827/8 and on Salisbury Plain in 1848/9. The measurement of the latter took nine months altogether. Some idea of the accuracy achieved is given by the fact that, assuming the Salisbury Plain base-line to be without error, a survey from it of the Lough Foyle base-line 350 miles away revealed a discrepancy of only five inches in its eight-mile length. The Ordance Survey proved itself to be of great use, not only for military purposes but to farmers, industrialists and planners.

Modern map-making

Although the outlines of most countries have now been carefully determined, it is nevertheless a fact that over 30% of the United States of America and more than 40% of the world has still to be accurately mapped. Indeed, the cartographic department of the United Nations is organizing a co-operative survey of the whole world with a view to producing an international map. Modern techniques include aerial surveys; the use of standardized radio signals has superseded accurate chronometers, while it is now possible to store geographical information in a computer and to select for print-out only specific information.

Two modern instruments are the geodimeter (Swedish) and the tellurometer (of South African origin), which is illustrated in Figure 98. The principle is the same in both. Using the geodimeter to find a large distance, OA, we have a light signal emitted from O and reflected back from A in a measured time. Knowing the speed of light, we can calculate

the distance OA. Of course, A may be a peak which is out of sight or obscured by cloud (see Figure 99). In this case a tellurometer is used. A radio signal from O is bounced back from a receiver/transmitter at A, and the time recorded. From the calculated distance OA and the measured elevation <AOB both the height of the mountain, AB, and its distance away, OB, may be determined.

Figure 99

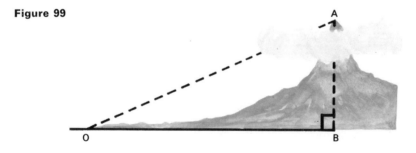

Over extensive, unhealthy or impassable terrain, such as mountainous country or rain forests, ground survey is often slow or impossible. In such cases aerial survey is used and, indeed, recent surveys show this as the most popular method. Aerial photography began in earnest during the 1914/18 World War for reconnaissance. In those days cameras were hand-held and there was inevitable difficulty in flying at a constant height and also in avoiding tilt so that a perfectly vertical and not

Figure 100 Aerial photograph of the Willesden
Railway Yard and Wembley Stadium (*Fairey Surveys*)

oblique photograph resulted. Later automatic levelling devices, fixed
camera mounts and automatic pilots were developed, along with
stereoscopic photography, in which overlapping two-dimensional
photographs were combined to reveal three-dimensional pictures of
ground features. Large, costly, stereo-plotting machines capable of
producing high-quality, contoured maps have also been developed.

Despite the problems of true height and distortion at the edges of
the photographs, the great advantage of aerial (and now satellite)
photography is, of course, its great speed, particularly in difficult and
fast-developing areas. In under-developed countries aerial photography
has played an invaluable part in economic planning, in deciding the
routes of roads and railways, the location of dams and power stations,
and even in estimating the agricultural potentiality of an area. In
Britain and other parts of Europe the method has been used in
archaeological research and the planning of highways and by-passes.

Mapping by computer

'In the past sixty years more information has been collected by man than during the whole previous period of human existence, and the next decade is likely to double this mass of knowledge.'

'It has been said that it is quicker to research a problem oneself than to search for the information in the literature.'

Both of these statements suggest that for many purposes, including mapping, data already exists. The problem is to store it in such a way that it is quickly accessible. Given that this can be done by storing the information on magnetic tapes, and that a computer can be programmed to select the data required without delay, there is still the problem of how to store a map outline simply as numbers. The method used is the obvious one of treating outlines, contour lines, and so on, as loci of large numbers of closely neighbouring points and giving their x and y co-ordinates relative to certain axes. In this way a map is stored as an immensely long list of x and y co-ordinates (or grid references). Kept on magnetic or paper tape or punched cards these constitute cartographic data banks. This method is being developed at the experimental cartography unit of the Royal College of Art in London and is financed by the Natural Environment Research Council. The first step is to convert existing maps into stores of numerical information. This process, called 'digitizing', is effected by a scanning device called a 'geameter', which follows outlines, contour lines, and so on, recording

Figure 101 For input the geameter scanner follows the lines automatically and converts them into a stream of digits (*A.E.G.*)

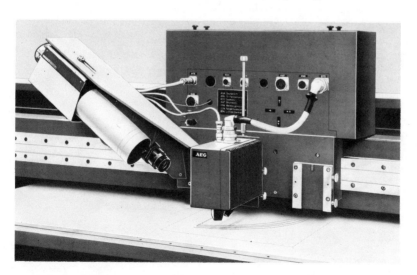

ten points every millimetre, and converts them into streams of number pairs. In the reverse or 'output' process the information is used to activate a scribing pen which draws out the information required. Speeds of plotting at present are just over six centimetres per second. To render the process economic this must be greatly increased.

Of greater interest still is the possibility of carrying out transformations of, or calculations upon, the stored data to produce other geographical statistics. For instance, where there exists a mathematical relationship between one map projection and another, the computer could obviously draw out, not the map stored, but some other projection of it. Then again, given the co-ordinates of two points (x_1, y_1) and (x_2, y_2), the distance between them can be calculated from the formula

$$\sqrt{\{(x_1 - x_2)^2 + (y_1 - y_2)^2\}}$$

So, by successive application of the formula, perimeters, coastline lengths, and so on, could be calculated. Alternatively, just as it is possible to calculate approximately the area enclosed by a curve defined either by co-ordinates or by an equation, so it should be possible to programme the computer to calculate land areas immediately without even producing the map.

Map projections

The only really true representation of the Earth is the global model and few geographers would be without a globe to hand. Unfortunately only very small scales are possible; a globe suitable for studying the county of Cambridge could scarcely be contained in a normal house! Moreover, globes are not only bulky but expensive to produce. The only alternative is somehow to convey geographical features from the Earth's curved surface to a plane one. There are a great variety of ways of doing this. They are called 'map projections'. Some are related by simple mathematical formulae; others are more complex. Some preserve correct shape and are called 'conformal' or 'orthomorphic'; some preserve correct area, but not shape and are known as 'equal-area' projections; and others do neither of these things. Some have been known since the days of Ptolemy; others are of far more recent origin. Three, which were known and used before the time of Christ, were projections of the sphere on to its polar tangent plane (see Figures 102–104). Let us imagine a translucent globe resting with its North Pole on a black, ground-glass screen with a small light bulb at L. When L is

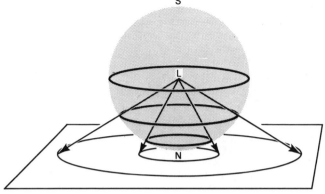

Figure 102 The polar zenithal or gnomonic projection

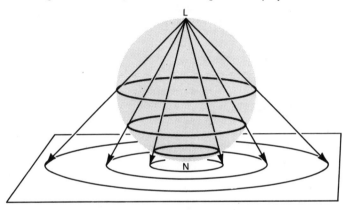

Figure 103 The stereographic projection

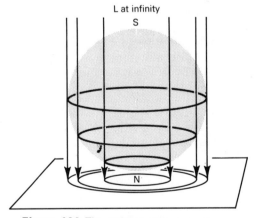

Figure 104 The orthographic projection

at the centre as in Figure 102, the projection is 'polar zenithal' or 'gnomonic'; when it is at the South Pole as in Figure 103, the projection is 'stereographic', and when L is at infinity and its light rays are all parallel and vertical as in Figure 104, the projection is 'orthographic'

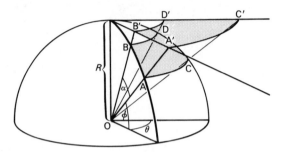

Figure 105 The polar zenithal projection of country ABDC

Let us look at these projections more closely. Figure 105 shows a small country, ABDC, and its polar zenithal projection A′B′D′C′. Now:

$$\frac{A'B'}{\text{arc } AB} = \frac{R[\cot \phi - \cot (\phi + \alpha)]}{R\alpha} = \frac{\sin \alpha}{\alpha \sin \phi \sin (\phi + \alpha)}$$

and

$$\frac{\text{arc } A'C'}{\text{arc } AC} = \frac{OA'}{OA} = \frac{R \operatorname{cosec} \phi}{R} = \operatorname{cosec} \phi$$

In this projection, circles of latitude project into concentric circles, and the scale, correct at the pole, increases outwards as the cosecant of the latitude and so becomes infinite at the Equator. Meridians of longitude obviously map into straight lines. If α is a small change of latitude, $\sin \alpha$ is approximately equal to α radians and

$$\frac{\sin \alpha}{\alpha \sin \phi \sin (\phi + \alpha)}$$

simplifies to $\operatorname{cosec}^2 \phi$. In this case, therefore, the scale increases outwards as the square of the cosecant. Now:

$$A'B' = AB \operatorname{cosec}^2 \phi$$

and

$$A'C' = AC \operatorname{cosec} \phi$$

Therefore we have:

$$\frac{A'B'}{AC} = \frac{AB \ \text{cosec} \ \phi}{AC}$$

For true shape to be retained under projection $\dfrac{A'B'}{A'C'}$ must equal $\dfrac{AB}{AC}$ and this is only true when cosec $\phi = 1$, that is, at and around the North Pole. Hence the projection is not conformal (or orthomorphic). Further:

$$\text{area} \ \frac{A'B'D'C'}{ABDC} = \frac{AB.AC \ \text{cosec}^3 \ \phi}{AB.AC}$$

$$= \text{cosec}^3 \ \phi$$

Again areas are not preserved except at the North Pole, so the projection is not an equal area one either.

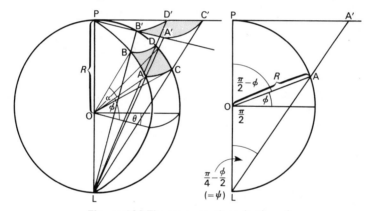

Figure 106 The stereograph projection of country ABDC

Figure 106 shows a stereographic projection from a point L at the opposite pole. In this case we have $\angle PLA = \pi/4 - \phi/2$. When ϕ increases to $\phi + \alpha$:

$$\angle PLB = \left(\frac{\pi}{4} - \frac{\phi}{2}\right) - \frac{\alpha}{2}$$

Thus:

$$A'B' = 2R\left[\tan\left(\frac{\pi}{4} - \frac{\phi}{2}\right) - \tan\left\{\left(\frac{\pi}{4} - \frac{\phi}{2}\right) - \frac{\alpha}{2}\right\}\right]$$

If we put $\pi/4 - \phi/2 = \psi$, then:

$$A'B' = 2R\left[\tan\psi - \tan\left(\psi - \frac{\alpha}{2}\right)\right]$$

$$= 2R\left[\tan\psi - \frac{\tan\psi - \tan\alpha/2}{1 + \tan\psi\tan\alpha/2}\right]$$

$$= 2R\left[\frac{\tan\psi + \tan^2\psi\tan\alpha/2 - \tan\psi + \tan\alpha/2}{1 + \tan\psi\tan\alpha/2}\right]$$

$$= \frac{2R\tan\alpha/2\sec^2\psi}{1 + \tan\psi\tan\alpha/2}$$

When α is small, $\tan\alpha/2 \simeq \alpha/2$, so:

$$A'B' = R\alpha\frac{\sec^2\psi}{1 + \tan\psi.\alpha/2}$$

But $AB = R\alpha$. Hence:

$$\frac{A'B'}{AB} = \frac{\sec^2\psi}{1 + \alpha/2\tan\psi}$$

Indeed, if α is very small, we can neglect it altogether and we have:

$$\frac{A'B'}{AB} = \sec^2\psi$$

But:

$$\frac{A'C'}{AC} = \frac{LA'}{LA} = \frac{2R\sec\psi}{2R\cos\psi} = \sec^2\psi$$

So, at any *point*, $A'B'/A'C' = AB/AC$; that is, the scale changes in projected lines of latitude and longitude are equal. Thus shape is preserved and the stereographic projection is conformal.

The third classical projection, the orthographic projection, produces the same distortions of scale found in aerial photographs. Here, projection is perpendicular to the target plane from a point infinitely distant. In Figure 107 we have:

$$A'B' = R[\cos\phi - \cos(\phi + \alpha)]$$
$$AB = R\alpha$$
$$A'C' = AC$$

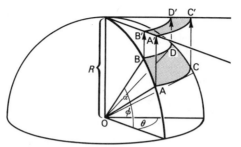

Figure 107 The orthographic projection of country ABDC

When α is small, $\cos \alpha \simeq 1$ and $\sin \alpha \simeq \alpha$ and A'B' reduces to $R\alpha \sin \phi$. Thus:

$$\frac{A'B'}{AB} = \sin \phi \quad \text{and} \quad \frac{A'C'}{AC} = 1$$

This projection is conformal only in the neighbourhood of the North Pole; elsewhere longitudinal distances diminish with the sine of the latitude.

The polar zenithal or gnomonic projection cannot be used to project the entire hemisphere whereas the other two projections can. All three project great circles into straight lines and are therefore useful for 'great circle sailing' in navigation. The stereographic projection is very useful for world maps which show two hemispheres; it is also employed by astronomers and navigators.

There are a number of variations of these projections, including a gnomonic one in which the scale along the meridian is progressively diminished so as to map equal areas onto equal areas, thus producing an equal area projection. Another, useful for mapping Equatorial regions, is obtained by projecting from the centre of the Earth onto a tangent plane at the Equator. With the point of contact (latitude 0°, longitude 0°), the projected lattice appears as shown in Figure 108. Meridians of longitude map into vertical straight lines, with the scale increasing outwards as the tangent of the longitude. Lines of latitude, however, map into a family of hyperbolae whose parametric equations are $x = R \tan \theta$, and $y = R \sec \theta \tan \phi$. For any given ϕ, we have:

$$\frac{y^2}{R^2 \tan^2 \phi} - \frac{x^2}{R^2} = 1 \tag{1}$$

since for all θ, $\sec^2 \theta - \tan^2 \theta = 1$. Equation (1) is the standard form of the equation of the hyperbola. In particular, latitude 45° ($\tan \phi = 1$)

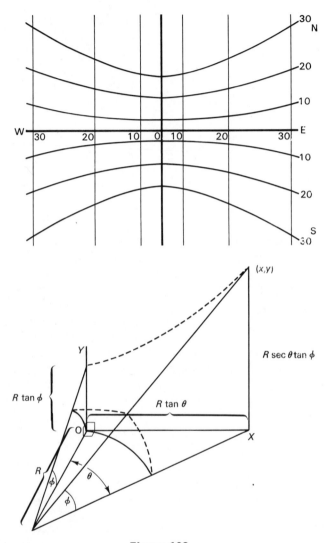

Figure 108

maps into a rectangular hyperbola. The projection obviously produces considerable distortion at longitudes and latitudes over 30°, but within these limits the projection is a useful one.

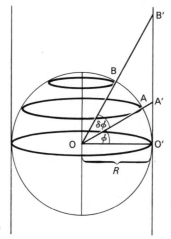

Figure 109

Later on, the spherical surface of the globe was mapped onto an osculating cylinder and cone. In the pure cylindrical projection illustrated in Figure 109, we have:

$$B'A' = R\left[\tan\left(\phi + \delta\phi\right) - \tan\phi\right)$$
$$= R\left[\frac{\tan\phi + \tan\delta\phi}{1 - \tan\phi\tan\delta\phi} - \tan\phi\right]$$
$$= R\left[\frac{\tan\phi + \tan\delta\phi - \tan\phi + \tan^2\phi\tan\delta\phi}{1 - \tan\phi\tan\delta\phi}\right]$$
$$= \frac{R\sec^2\phi\tan\delta\phi}{1 - \tan\phi\tan\delta\phi}$$

and if $\delta\phi$ is very small

$$\simeq R\sec^2\phi\,\delta\phi$$

Also $BA = R\,\delta\phi$, and therefore:

$$\frac{B'A'}{BA} = \sec^2\phi$$

Hence the scale along a meridian increases as the square of the secant of the latitude. For the latitudes we have:

$$\frac{\text{latitude through A}'}{\text{latitude through A}} = \frac{OA'}{OA} = \frac{R\sec\phi}{R} = \sec\phi$$

This means that the scale along the latitude increases as the secant of the latitude. The pure cylindrical projection is thus neither equal area nor orthomorphic, and it is only useful for Equatorial regions. However, two variants are possible. First, if the distance between the lines of latitude obtained by pure cylindrical projection is scaled down in the ratio sec ϕ : 1, an equal area projection is obtained. Distortion at high latitudes is excessive and so this variant is usually only employed for tropical latitudes. Secondly, there is the famous *Mercator projection*, in which the pure cylindrical projection is so modified that the scale along the meridians is made to increase in exactly the same ratio (that is, sec ϕ : 1) as it increases along the parallels. This makes the projection orthomorphic. More important, rhumb-lines (that is, lines of constant course) are projected as straight lines; this is the chief value of the projection. It is not, of course, equal area. In fact, the scale of areas increases in the ratio of the square of the secant of the latitude. Thus, one characteristic of Mercator's world map is an immense Greenland and a tiny India. At the Equator sec² ϕ = 1 and arcs are 'true', but at latitudes ϕ = 45°, 60°, and 75°, for example, sec² ϕ = 2, 4, and 14.9 respectively. Thus, if Equatorial areas are 'true', those on latitudes 75° North and South are enlarged nearly fifteen times:

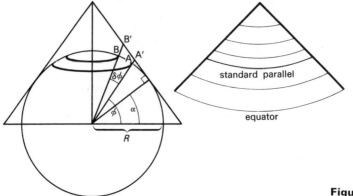

Figure 110

The simple, perspective, conical projection illustrated in Figure 110 is seldom used. The scale along a meridian increases with the latitude. At latitude ϕ, for example, we have:

$$\frac{B'A'}{BA} = \frac{R[\tan (\phi + \delta\phi - \alpha) - \tan (\phi - \alpha)]}{R \, \delta\phi} \simeq \sec^2 (\phi - \alpha)$$

When $\phi - \alpha = 0$, $\sec^2(\phi - \alpha) = 1$ and the scale is true. This is obviously so, for the parallel is in actual contact with its map. We call this the *standard parallel*. However, the projection is neither equal area nor orthomorphic and is quite useless for high latitudes. A modification of the perspective conical projection is to shrink the conical 'cap' until it cuts the Earth along *two* standard parallels (see Figure 111). At

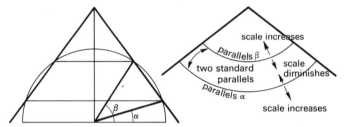

Figure 111

each of these the scale is correct; between them it diminishes; outside it increases. This projection is used for mapping Europe, Asia, North America, and Australia.

In the 'conventional', simple conical projection, a non-perspective construction, the spaces between parallels of latitude are adjusted to make the scale along every meridian correct; that is, equispaced parallels of latitude are drawn as equispaced, concentric circular arcs on the projection. It is rather as though the arc BA were 'unwound' until it lay

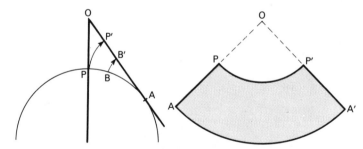

Figure 112

along B'A' on the cone's surface (see Figure 112). Consequently, the pole P maps onto a whole circular arc of radius OP' and the projection is an incomplete sector. The projection is sometimes used for mapping the Mediterranean Sea.

A very common 'equal area' modification of the conical projection is the Bonne projection. In this, a parallel (say, ϕ N) is selected as standard and is drawn in the usual way as a circular arc of length $2\pi R \cos \phi$ and radius $R \cot \phi$. All other equally spaced parallels are then drawn as equally spaced, circular arcs having the same curvature as the standard. Each parallel is then divided correctly and the corresponding points of division are joined to give the meridians. Figure 113

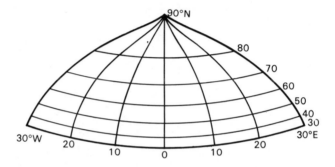

Figure 113 Bonne projection of region between latitudes 30° and 90°N

shows a Bonne projection of the region between latitudes 30° and 90° North with standard parallel 60° N and bounded by the meridians 30° W and 30° E.

A special case arises when the above procedure is used for the whole range of latitude and longitude, with the Equator as the standard parallel and the longitude 0° as central meridian (see Figure 114).

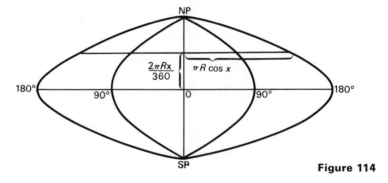

Figure 114

Clearly, the Equator, and therefore all other equispaced latitudes, map into parallel and equispaced straight lines. The length of latitude $x°$ N is $2\pi R \cos x$ and so the profile of each meridian is a cosine curve with O as origin, or a sine curve with the North Pole as origin. It is known as the *Sanson–Flamsteed sinusoidal projection* and is frequently employed for world maps.

The only other projection we will look at here is the Mollweide or homolographic projection. It is a world-map projection in which parallels of latitude are projected, as in the sinusoidal projection, into parallel straight lines, while the meridians are drawn as semi-ellipses. The meridian 180° E is represented by a semi-ellipse of semi-major axis $2R$ and semi-minor axis R. The construction is so carried out that the mapped area between the mapped Equator and the mapped parallel ϕ is equal to the area of the spherical zone between the same parallels on the actual globe projected.

There are many more projections, including types such as polyconics and interrupted projections, and special constructions such as the transverse Mercator, the Eckert IV, and the Cassini projection used in the one-inch Ordnance Survey of Great Britain with a fixed straight reference meridian through Delamere Forest in Cheshire. They are described in the specialized literature.

Selenography—mapping the moon

The study of the Moon's surface began with Galileo in 1609. It is almost unbelievable that a year before John Speed published his ornate, now highly prized, but nevertheless inaccurate maps of the English counties —many of them simply decorated versions of Saxton's first ever county maps—Galileo had already turned his attention towards the Moon. With the aid of the reflecting telescope—his own invention—he saw shadows cast by mountains, dark plains and several of the large craters. Looking closely at the boundary between the lit and unlit regions he noticed, just within the darkened areas, bright pin-points, and he realized that these were mountain peaks! Not only that, but by gauging their distances from the 'terminator' (the dark boundary) when they appeared or disappeared, he was able to estimate their heights. His achievements are commemorated in the name of a tiny crater (located in the region numbered 15 in Figure 115).

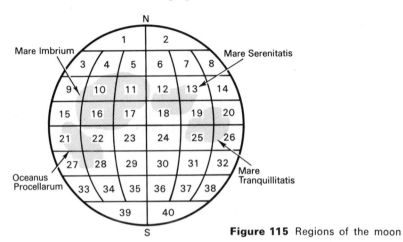

Figure 115 Regions of the moon

The first maps of the Moon appeared between 1620 and 1640. They were devised by Langrenus, a Belgian who was cartographer to the King of Spain. Although his name too is commemorated in the name of a crater—a large one in region 26 of the visible side of the Moon, his work received little attention at the time. Of greater note is the 1647

Figure 116 Langrenus crater photographed from *Apollo* 8 at an altitude of 150 nautical miles (*U.S.I.S.*)

map of Johannes Helvelius, a brewer, city councillor of Danzig and amateur astronomer.

The modern system of naming the lunar plains and craters dates from 1651 when Riccioli of Bologna, a Jesuit, published a map devised by his pupil Grimaldi. This showed the Mare Serenitatis (Sea of Serenity), the Mare Imbrium (Sea of Showers), and the Oceanus Procellarum (Ocean of Tempests). Other features were named after philosophers and men of science—Plato (in region 5 of Figure 115), Ptolemy (23), Copernicus (16, 17), Tycho (35), and so on. Since then, as any modern map of the Moon will show, other craters have been named after such celebrated mathematicians as Apollonius, Archimedes, Babbage, Barrow, Bernouilli, Cassini, Cauchy, Cayley, Clairaut, D'Alembert, Descartes, Eratosthenes, Eudoxus, Euclides, Euler, . Fermat, Fourier, Gauss, Jacobi, Klein, Lagrange, Legendre, Maclaurin, Menelaus, Newton, Pythagoras, Recorde and Taylor. Astronomers and other scientists have their memorials in the craters Cavendish, Celsius, Daniell, Eddington, Faraday, Flamsteed, Franklin, Foucault, Fraunhofer, Fresnel, Herschel, Hooke, Huyghens, Janssen, Kelvin, Kepler, Lavoisier, Leibnitz, Oersted, Posidonius, Ramsden, Rutherford, Toricelli and Watt. Vasco da Gama and Julius Caesar are both remembered, as well as many other famous men.

In 1775 T. Mayer of Göttingham produced a map whose accuracy was not surpassed until 1824. In 1779 J. H. Schröter, who worked with William Herschel, began calculations to prove that the Moon was inhabited. Most of the maps up to this time were based on careful drawings of telescopic images, but just as the vernier and micrometer, and an improved theodolite became available to Picard and Roy in their terrestrial surveys, so were the astronomers able to improve the accuracy of their work. Between 1824 and 1836, W. Lohrman of Dresden achieved greater accuracy by fitting a micrometer to the eye-piece of his telescope. His survey resulted in a fifteen-inch diameter, full Moon map, which was published in 1838. Subsequently, this survey was extended by Bernhard Schmidt who, in 1878, enlarged and engraved the map in twenty-five sections, which showed altogether 7178 craters and 99 clefts.

In 1837 W. Beer and J. H. Mädler had published their illustrated *Der Mond*, which presented the Moon as an unchanging world. However, great excitement was generated in 1868 when Schmidt was unable to discover a small crater which Mädler had called Linné; the implication that the Moon's surface was not static and unchanging inspired a great many amateur astronomers to keep a close watch upon it. A Selenographical Society flourished between 1878 and 1890, when the British Association set up a Lunar Section with T. G. Elgar as its first director.

The modern way of producing Moon maps is to use measurements from aerial photographs taken from satellites and space probes. Using this method, for instance, J. H. Franz and S. A. Saunders determined the positions of 3000 additional objects to an accuracy of 1000 feet. These were used as standard points. In 1960 D. W. G. Arthur and E. A. Whitaker produced the most reliable orthographic Moon atlas so far, and they revised it in 1964. This used a new co-ordinate standard grid based on several thousand standard points. In recent years in the US, the National Aeronautics and Space Administration (NASA) has, through photographs obtained from its series of *Orbiter* satellites and, later, its series of *Apollo* Moon shots, accumulated a vast quantity of detailed information, not only about the physical features of the visible and invisible sides of the Moon, but also about atmospheric conditions and other physical phenomena associated with the Moon's immediate environment. During the same period the Soviet Union has been collecting similar information from its own Moon shots.

The first really spectacular close-up photographs of the Moon were obtained by the crew of *Apollo* 8, Frank Borman, William Anders and James Lovell, as they orbited the satellite at Christmas 1968. They brought back thousands of close-range shots of the near and far side regions, together with some spectacular views of the Moon's horizon with the Earth just rising above it. *Apollo* 9 was concerned with practising, in orbit, transposition and docking procedures for the lunar module *Moonbug*. Then *Apollo* 10 repeated these manœuvres round the

Figure 117 The earth seen rising slowly above the moon's horizon (*U.S.I.S.*)

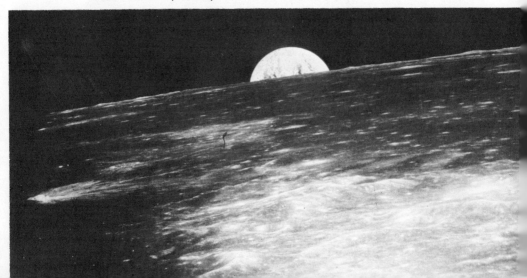

Figure 118 The second man on the moon—Buzz
Aldrin, a crew member of *Apollo* 11 (*U.S.I.S.*)

Moon itself, with the parent space craft in a circular orbit sixty-nine
miles (111 km) above the surface, while the module orbited independently to within ten miles (16 km) of the surface, later redocking with
the parent craft.

To the crew of *Apollo* 11, Neil Armstrong, Michael Collins and Buzz
Aldrin, fell the honour of making the first manned landing on the Moon.

Lift-off was on 16 July 1969 from Cocoa Beach, Florida. Seventy-five hours later the vicinity of the Moon was reached, and 109 hours 51 minutes after lift-off two men, Armstrong and Aldrin, had their feet on the lunar surface. It was an historic achievement, and the whole enterprise was impeccably executed. The crew returned with an immense number of scientific readings and photographic records, lunar rocks and Moon dust. They set up and left several instruments including a TV camera, seismometer and a mirror device for reflecting laser beams.

Apollo 12, manned by Charles Conrad, Richard Gordon and Alan Bean, orbited the Moon between 14 and 24 November 1969. On this occasion two men between them spent a total of 14 hours 43 minutes on the Moon's surface. Although this expedition received far less publicity than did *Apollo* 11, it achieved as much, if not more, in terms of solid scientific work. The seismometer set up then has been radioing back monthly signals to Earth. These have shown monthly Moon quakes when nearest the Earth, the strongest originating from the Fra Mauro region—site for the successful *Apollo* 14 landing—and doubly intriguing scientists, who believed the rocky rubble of that region to date from the Moon's formation 4600 million years ago!

The ill-starred *Apollo* 13 Moon shot in April 1970 was aborted after an explosion, fifty-six hours into the mission, in the service module. By an incredible combination of ingenuity and courage the crew, James Lovell, Fred Haise and John Swigert, cooped up in the lunar module, blasted themselves into a 'super-fast' return trajectory and to the great relief of waiting millions splashed down safely on 17 April, $3\frac{1}{2}$ miles (5.6 km) from the recovery ship.

The necessary *post mortem* caused alterations in the *Apollo* programme, which delayed the *Apollo* 14 right until 31 January 1971. Manned by Alan Shepherd, Edgar Mitchell and Stuart Roosa, one of the main objects of this flight was to discover whether water is locked in the Moon's rocks. After *Apollo* 15 in July 1971 and *Apollo* 16's successful flight in April 1972, *Apollo* 17 is the last scheduled American moon launch.

Meanwhile, the next space 'first' was scored by the Soviet Union with the successful landing at the end of November 1970 of their revolutionary moon-crawler, *Lunokhod* I. Remote-controlled by a crew of five on Earth and solar-powered, the vehicle parked near its mother ship at night and travelled by day, carrying out a wide range of experiments. It obtained far more high-quality lunar surface close-ups than did the Russian *Luna* 9 and 13 together and, using an X-ray spectrometer, analysed the lunar surface at intervals of several dozen metres. It coped successfully with several little hills and valleys, rocks and craters, and it recorded quite a variety of surface textures from hard and rocky to soft and powdery. A similar US mission was successfully carried out in July 1971. Unhappily, the Russian *Soyuz* III expedition which also took place in July 1971, ended in the tragic death of its three astronauts as their spacecraft returned to earth.

The most spectacular, unmanned probes of the 1970s announced so far are the NASA 'grand tours' of the solar system. The first, in 1977, may circuit Jupiter and Uranus and end on Neptune in 1988. The second, scheduled for 1979, is via Jupiter and Saturn to Pluto. Nearer home, short, unmanned 'Explorer' trips to Jupiter, Venus, Mercury and Mars are time-tabled for 1972, 1973, and 1976.

By way of summary here are a few lunar statistics.

(1) The Moon orbits elliptically with the Earth's centre as one focus. Its distance from the Earth thus varies between 356 000 and 407 000 kilometres or, in miles, to the nearest thousand, 221 000 to 253 000. Thus, the average distance is 239 000 miles and the eccentricity of the almost circular orbit roughly 0.055.

(2) The Moon, like the Earth, turns upon its own axis, but it does this once in exactly the same time as it takes to circle the Earth—$27\frac{1}{3}$ Earth days. For this reason the same side of the Moon is always facing us. However, the rotation axis of the Moon is not quite perpendicular to the plane of its elliptical path, and so it is not always the same point on the Moon's surface that is nearest to the Earth. In fact, during a month nearly 60% of the Moon's surface (though never more than 50% at any one time) is visible from the Earth.

(3) The diameter is 3476 kilometres or 2160 miles, and the mean density 3.36 times that of water. (The Earth's mean density is 5.5 g/cm^3.) Consequently its volume is about $\frac{1}{50}$ of the Earth's and its mass less than $\frac{1}{80}$. The marked difference in density between the Earth and its satellite indicates quite different chemical compositions and does not support the idea that both were once one body. The Moon's lesser mass produces a pull of gravity only $\frac{1}{6}$ of that of the Earth's. This means that a heavy, 100-kilogramme man tips the scales on the Moon at a spare 17 kilogrammes—hence the need for leaden boots like those of a deep-sea diver! But the Moon's lower mass also means a lower escape velocity: 1.5 miles a second (2.4 km/sec) compared to over 5 miles a second on Earth. It also means that water or gas molecules have probably long since deserted the satellite. As a result, the well-known

'greenhouse' effect—by which carbon dioxide, in particular, and other gases let heat through but not out again, just as in a glass greenhouse— no longer functions; hence the violent fluctuations of surface tempera- ture recorded, from 100°C in the daytime to −150°C at night. Evening walks on the Moon could be chilly as well as eerie experiences!

Shapes and Sizes in Geography

'Multiplication is vexation; Division is bad; The Rule of three doth puzzle me, And Practice drives me mad.'
Elizabethan MS dated 1570

'The secret of the arts is to correct nature.'   VOLTAIRE

'The reasonable man adapts himself to the world; the un-reasonable one persists in trying to adapt the world to himself. Therefore all progress depends on the unreasonable man.'   GEORGE BERNARD SHAW

Of all the branches of pure mathematics currently described as 'new' or 'modern', topology alone can claim to have been developed during the present century; it is, moreover, still growing vigorously. Starting with Euler's 'Bridges of Königsberg' puzzle and Möbius' five-colour problem, it has now grown into a far-ranging and immensely powerful branch of mathematics. It has been used to simplify and generalize whole areas of classical analysis. Applied, it has yielded solutions to previously insoluble industrial problems. In this context, however, it is probably most helpful to think of topology as the most general kind of geometry there is. Let us imagine all the straight lines, triangles, quadrilaterals, circles, and so on, of our schooldays drawn on thin rubber sheeting which can be stretched or deformed (but not torn) in any way desired. For instance, we might draw a shape on a balloon and then blow it up. Topology asks questions about the shape drawn and the shape it becomes after inflation. Angles, lines, shape and area have all changed. Has anything remained constant? Is there anything worth saying at all? The only obvious invariant is that neighbouring points on the original shape are still neighbours after inflation. Had we marked three points, A, B, and C in that order, on an arrowed line in the first place, they would still be traceable in the same order after inflation.

Let us consider a spherical surface. This can be deformed, without tearing, into a host of other shapes—a cube, a pancake, a sausage, and so on. In every case neighbouring points remain neighbours, and we say the surfaces are topologically equivalent. For one thing, a simple knife cut will sever each shape in two (see Figure 119(a)). For another,

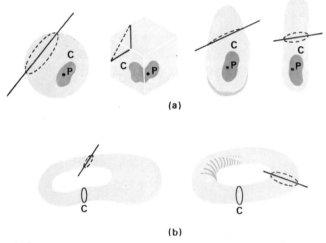

(a)

(b)

Figure 119

any closed curve, C, on the surface may be shrunk continuously to a point P. Neither of these facts is true of an inflated rubber ring, or torus, which can only be formed from the sphere by poking or tearing a hole in it (see Figure 119(b)). So, in these respects at least, the torus has different properties from the sphere.

Let us now look at the old puzzle of joining houses A, B, and C to electricity (E), water (W), and gas (G) supplies. At least one set of supply lines must cross (see Figure 120). But what if the world was not spherical but torus-shaped? Figure 121 shows that in this case the supply lines do not necessarily cross.

The eighteenth-century Swiss mathematician, Leonhard Euler, showed that for a spherical surface divided up into F regions by E boundaries meeting at V nodes (or corners, that is, points where three or more edges meet), then $V - E + F$ was always equal to 2. In Figure 122,

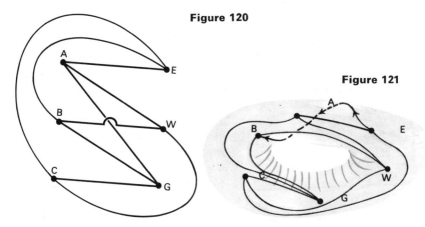

Figure 120

Figure 121

$V = 2$, $E = 3$, and $F = 3$. Hence $V - E + F = 2$. However, on the torus with one hole, Euler found that the result of this expression was zero $(2 - 2 \times 1)$, while on the p-holed torus $V - E + F = 2 - 2p$. In Figure 123, $V = 1$, $E = 2$, and $F = 1$. Hence $V - E + F = 2 - 2 \times 1 = 0$.

Figure 122   Figure 123

The nineteenth-century German mathematician, August Möbius, pondered the problem of finding the least number of colours required on a map to distinguish political regions, or countries, or counties, given that each boundary line should separate two differently coloured regions. Figure 124 shows three 'maps' requiring at least two, three and

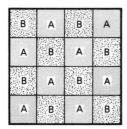
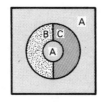

Figure 124

four colours respectively. Möbius set his students the problem of finding a map which required at least five colours. As it turned out, the students were unsuccessful. Nor has anyone else since been able to draw such a map. On the other hand, no mathematician has yet proved that four colours will always suffice! Nevertheless, some progress has been made. It has been proved that any map with not more than thirty-eight regions can always be coloured with four colours; it is also certain that on a plane, or spherical surface, no six-colour map can ever be discovered.

On the torus, however, the problem simply vanishes. Five-colour maps are trivially easy to draw. In fact it is not very difficult to discover a map which requires seven colours. Imagine a piece of rubber sheeting divided as shown in Figure 125 and rolled so that the edge DC lies along and becomes fastened to AB, and then bent round so that the edge AD—now forming a closed shape—joins up with the edge BC—similarly a closed shape. We then have a world shaped like a torus,

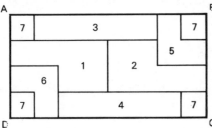

Figure 125

divided into countries in such a way that seven different colours are needed to distinguish the boundaries. We recall that for an *n*-holed torus (or a surface of *genus n*) $V - E + F = 2 - 2n$. P. J. Heawood has shown that for all surfaces where $n \geqslant 1$, then the sufficient number of colours is given by the expression:

$$\left[\frac{7 + \sqrt{(1 + 48n)}}{2} \right]$$

where [*n*] is the greatest integer not exceeding *n*. For the one-holed torus this asserts that the *chromatic number*, or the sufficient number of colours, is:

$$\left[7 + \frac{\sqrt{(1 + 48 \times 1)}}{2} \right] = [7] = 7$$

Figure 126

So our seven-colour map is the best we can do on a rubber ring. What about a world with two holes (see Figure 126)? Heawood's formula gives the chromatic number as:

$$\left[\frac{7 + \sqrt{97}}{2}\right] = [8.42] = 8$$

The reader may care to investigate further. It is, for instance, an interesting exercise to search a political atlas for continents which require four colours.

The rubber sheet geometry idea is sometimes used in geographical maps where 'true scale' lengths and areas are not needed. The most common example occurs in the London Transport underground railway maps, where all that matters is that the sequence of stations on each

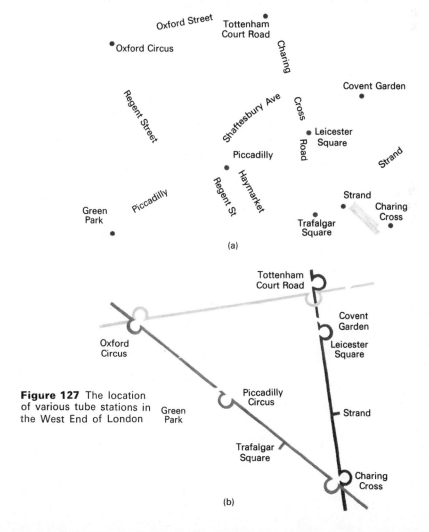

(a)

Figure 127 The location of various tube stations in the West End of London

(b)

The map shows 95 administrative counties of Scotland, England and Wales

The map is correct topologically since contiguity of counties has been preserved

The area of each county is proportional to the population of the county in 1961

□ 10 000 people
⊞ 40 000 people

Key

1.	London (old county)	33.	Suffolk (E)
2.	Middlesex	34.	Norfolk
3.	Essex	35.	Holland
4.	Kent	36.	Kesteven
5.	Surrey	37.	Lindsey
6.	East Sussex	38.	Nottinghamshire
7.	West Sussex	39.	Derbyshire
8.	Hampshire	40.	Staffordshire
9.	Isle of Wight	41.	Cheshire
10.	Berkshire	42.	Lancashire
11.	Oxfordshire	43.	West Yorkshire
12.	Buckinghamshire	44.	East Yorkshire
13.	Bedford	45.	North Yorkshire
14.	Hertfordshire	46.	Durham
15.	Cornwall	47.	Northumberland
16.	Devon	48.	Cumberland
17.	Dorset	49.	Westmorland
18.	Somerset	50.	Monmouth
19.	Wiltshire	51.	Glamorgan
20.	Gloucester	52.	Pembroke
21.	Herefordshire	53.	Carmarthen
22.	Worcestershire	54.	Cardigan
23	Shropshire	55.	Brecknock
24.	Warwickshire	56.	Radnor
25.	Northamptonshire	57.	Montgomery
26.	Leicestershire	58.	Merioneth
27.	Rutland	59.	Caernarvon
28.	Soke of P.	60.	Anglesey
29.	Huntingdonshire	61.	Denbigh
30.	Isle of Ely	62.	Flint
31.	Cambridgeshire	63.	Wigtown
32.	Suffolk (W)	64.	Kirkcudbright

65.	Dumfries
66.	Roxburgh
67.	Selkirk
68.	Peebles
69.	Berwick
70.	East Lothian
71.	Midlothian
72.	West Lothian
73.	Fife
74.	Kinross
75.	Clackmannan
76.	Stirling
77.	Dunbarton
78.	Renfrew
79.	Lanark
80.	Ayr
81.	Bute
82.	Perth
83.	Angus
84.	Kincardine
85.	Aberdeen
86.	Banff
87.	Moray
88.	Nairn
89.	Shetland
90.	Orkney
91.	Caithness
92.	Sutherland
93.	Ross and C.
94.	Inverness
95.	Argyll

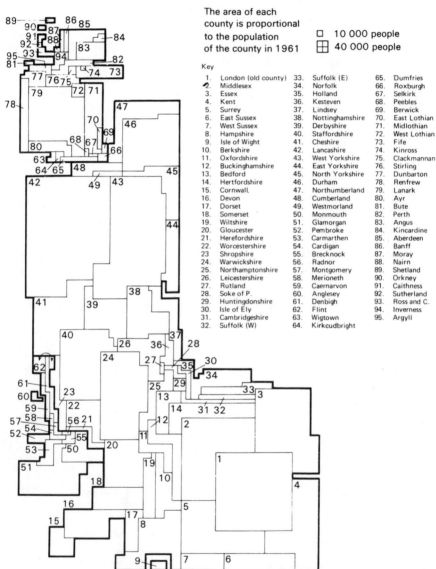

Figure 128 A population map

line is correct. Figure 127(a) is a 'true scale' geographical map of a well-known part of the West End of London showing the locations of various tube stations; Figure 127(b) is a topological transformation, or 'rubber sheet stretch', where only the manner in which the stations are *connected* is shown.

On other occasions we may well wish to illustrate not the actual sizes and shapes of, say, the various English counties, but, for example, the population of each. J. P. Cole of Nottingham University, who is well known for his research into the application of modern mathematical techniques in geography, has produced the topological maps shown below (see Figures 128–130). Figure 128 is a topological transformation of the usual map of Britain; the area of each county on the map is not proportional to its actual physical area and shape but to its population. The same device may be employed to give a picture of the variation of any statistic over the country. Figure 129 shows a topological transformation of England and Wales in which the areas of the various counties are proportional to the frequency of their mention in the *Eastern Daily Press*. It is interesting to note that at one time in our history our population map would also have resembled this shape.

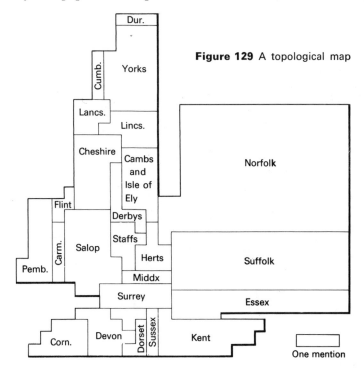

Figure 129 A topological map

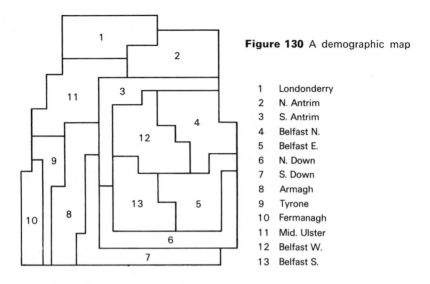

Figure 130 A demographic map

1	Londonderry
2	N. Antrim
3	S. Antrim
4	Belfast N.
5	Belfast E.
6	N. Down
7	S. Down
8	Armagh
9	Tyrone
10	Fermanagh
11	Mid. Ulster
12	Belfast W.
13	Belfast S.

Figure 130 is a demographic map showing electoral strength in Northern Ireland.

Lorenz curves

The demographic population graph in Figure 128 enables us to see at a glance the heavily and sparsely populated areas, but we can analyse the distribution more carefully. Suppose we wish to investigate the variation of population density over, say, the north midland counties of England. The population figures for 1961 are tabulated below. From these, and the known acreage, we arrange the counties in descending order of population density, calculate areas and populations as percentages of the whole, and finally express them as cumulative percentages.

North Midland counties	population (1000)	area (1000 acres)	population density (persons per acre)	area as % of whole	population as % of whole
Notts.	903	540	1.7	13	25
Derby.	878	644	1.4	16	24
Leics.	683	532	1.3	13	19
Northants.	398	585	0.7	14	11
Lincs.	744	1704	0.4	42	20
Rutland	24	97	0.2	2	1

cumulative percentages	
area	population
13	25
29	49
42	68
56	79
98	99
100	100

By plotting these values we obtain the Lorenz curve as shown in Figure 131(b). If the density of population were exactly the same in each county, the Lorenz curve would simply be the straight line equally

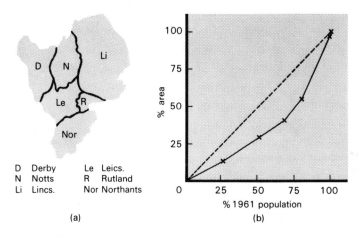

D Derby	Le Leics.
N Notts	R Rutland
Li Lincs.	Nor Northants

(a)

(b)

Figure 131 Lorenz curves showing population density in the North Midlands, 1961

inclined to both axes. In fact, it can be shown that the area between the two is a useful measure of the spread of population density over the whole region considered.

We can also use the Lorenz curve to point out changes or drifts of population over a period of time. An example of this is provided by the population figures for the three Ridings of Yorkshire in 1801 and 1961. Before examining Figure 132, let us look at the table on the next page.

Yorkshire	area (1000 acres)	population 1801 (1000)	1961 (1000)
West Riding	1786	591	3645
East Riding	750	111	527
North Riding	1362	158	554

	area as % of whole	pop. as % of whole 1801	1961
West Riding	46	69	77
East Riding	19	13	11
North Riding	35	18	12

cumulative percentages

area	population 1801	1961
46	69	77
65	82	88
100	100	100

The Lorenz curves for 1961 and 1801 show a marked divergence from the broken line for uniform density population (see Figure 132(b)).

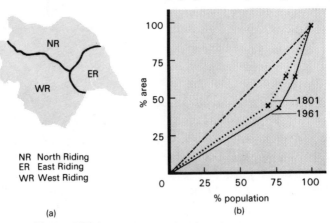

NR North Riding
ER East Riding
WR West Riding

(a)

(b)

Figure 132 Lorenz curves showing changes in population density from 1801 to 1961 in Yorkshire

Clearly, the trend is explained by the drift from the countryside to the West Riding industrial towns during the Industrial Revolution. The Lorenz curve can, of course, be used to show the distribution of occupations, plant species, and a variety of other statistical data.

Topology and trees

We have seen in Chapter 6 some situations in economics in which closed networks arise. Frequently, situations must be represented by open networks. The most common is the family tree. Part of the interesting but incomplete one of Christopher Saxton is shown below.

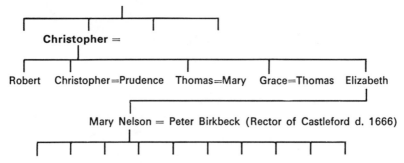

The line of royal succession from Tudor times to the present day is also a kind of 'critical path' along a tree diagram, and so is the hierarchical structure of the chain of command in an army or business organization or Ministry of the Crown. In the world of computers the flow chart is another kind of tree, though unlike the previous examples it is apt to contain closed loops between its beginning and end.

In geography the obvious examples of 'trees' are provided by rivers and their tributaries. A topologist seeks to analyse trees, and even to set up criteria of topological equivalence. Using the same notation as before, we denote the number of lines, or segments, in the tree by E, the number of points at which one line ends by V_1, the number of points at which three lines meet by V_3, the meeting of four lines by V_4, and so on. In this scheme V_2 would apply to every other point on the tree network, and this would make $V_2 = \infty$ (infinity). To avoid this we argue instead that every point $\wedge$ where two lines *obviously* meet can be topologically transformed (by straightening) into one continuous line; then $V_2 = 0$. Finally, we write:

$$V = V_1 + V_3 + V_4 \ldots$$

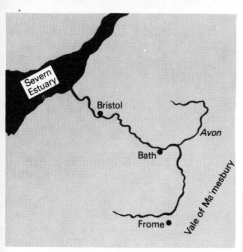

Figure 133 The Bristol *Avon*

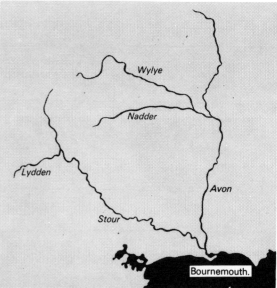

Figure 134 The Bournemouth *Avon*

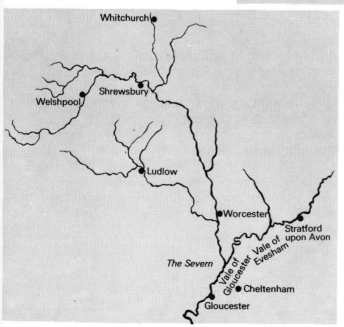

Figure 135 The *Severn*

Figures 133, 134 and 135 show respectively the Bristol *Avon*, the Bournemouth *Avon* and the *Severn*. Topologically they can be simplified as follows:

Bristol *Avon*

Figure 136

$E = 3$; $V_1 = 3$, $V_3 = 1$; $V = V_1 + V_3 = 4$, and $V - E = 1$.

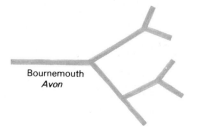

Bournemouth
Avon

Figure 137

$E = 9$; $V_1 = 6$, $V_3 = 4$; $V = V_1 + V_3 = 10$, and $V - E = 1$.

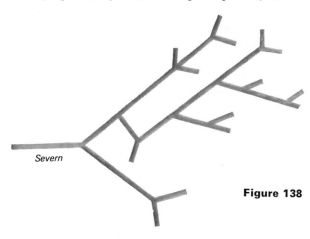

Severn

Figure 138

$E = 26$; $V_1 = 15$, $V_3 = 11$, $V_4 = 1$; $V = V_1 + V_3 + V_4 = 27$, and $V - E = 1$.

Q

If we were to put $F = 1$ in each case, we should have $V - E + F = 2$, the familiar Euler's theorem, though, of course, it would not hold good for rivers on the torus!

Now, in each case we work out: $1 \times V_1 + 2 \times V_2 + 3 \times V_3 + 4 \times V_4$, and so on ($V_2 = 0$). Then for the Bristol *Avon* we have:

$$1 \times 3 + 2 \times 0 + 3 \times 1 = 6 = 2 \times E$$

For the Bournemouth *Avon* we have:

$$1 \times 6 + 2 \times 0 + 3 \times 4 = 18 = 2 \times 9 = 2 \times E$$

And for the *Severn* we have:

$$1 \times 15 + 2 \times 0 + 3 \times 11 + 4 \times 1 = 52 = 2 \times 26 = 2E$$

It looks as though for trees in general:

$$2E = V_1 + 2V_2 + 3V_3 + 4V_4 \ldots nV_n$$

Of course, our three cases do not really prove the result any more than 300 special cases would, but it makes us begin to suspect patterns in trees, and to want to establish or refute our hunches by proof.

To establish criteria for topological equivalence of trees or rivers, we must introduce more terms. The centre of a tree is reached by removing branches ending in end-points or 'V_1' points. So, in Figure 139, we have:

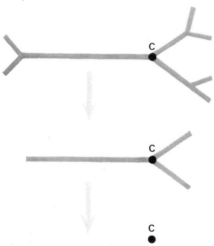

C

C

C

Figure 139

C is called the *centre* of the tree.

However, Figure 140 gives us:

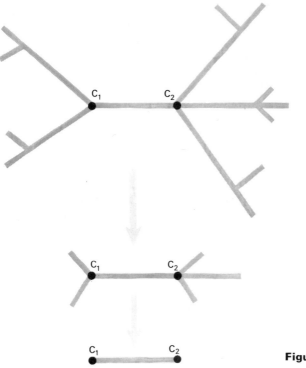

Figure 140

And we call $C_1 C_2$ a *bicentre*.

The *path length* between two points of a tree is the number of branches or line segments separating them. Thus in Figure 141 the path from A to F is of length 2. But we can find paths longer than this, and the longest one possible is called a *diameter*. However, as with the circle, there may be more than one diameter. In this case the diameter is of length 4 and there are six diameters—ADFGH, BDFGH, CDFGH, ADFGJ, BDFGJ, and CDFGJ. We should also notice that every diameter passes through the centre F!

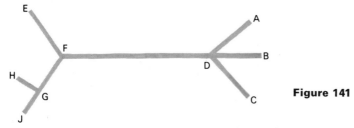

Figure 141

The reader may care to analyse some rivers—finding their centres and diameters. Writing in *Mathematical Education in Science and Technology*, Professor C. A. M. King points out that for river systems and drainage networks generally with t outer tips or stream sources, b stream confluences, r outlets and E links between nodes, we have $E = b + r + t - 1$ and $b + r = t$. So for the Bournemouth *Avon*, $t = 5, r = 1, b = 4$, and $E = 9$—values which check in the formulae above.

Shape and compactness

Figure 142 shows the outlines of various countries ranging from long, thin, attenuated Chile, through the fragmented Philippines, to chunky, compact Uruguay. Obviously Norway is more compact than Chile but less so than England and Wales. But how do we *measure* this compactness so that we may distinguish between, say, the compactness of

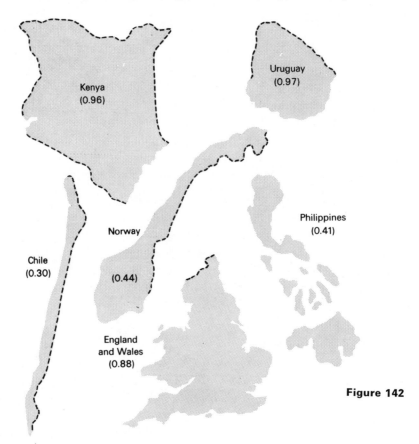

Figure 142

Kenya and Spain, or of other countries where visual judgement cannot discriminate? A most interesting paper on the whole subject, with a full mathematical treatment, was produced by D. J. Blair and T. H. Bliss (as one of a series of bulletins produced under the auspices of the Geography Department of Nottingham University in 1967).

Various definitions of compactness have been suggested. N. J. G. Pounds used the numerical ratio of boundary length to area, but this yields widely different results for the figures shown below, their areas being alike but perimeters very different (see Figure 143). Yet reason

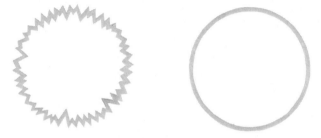

Figure 143

suggests that they are (roughly) equally compact. V. C. Millar's definition of the compactness of a shape as the ratio of its area to that of a circle having an equal perimeter suffers from a similar disadvantage; the results for Figure 118 would be just as misleading by this method.

Cole defines compactness of a figure as the ratio of its area to the area of the smallest circle that circumscribes it. This works well in many cases but would give an unreliable result in a case such as that shown in Figure 144, where everything depends on the freak peninsula, though it is a very small fraction of the total area and the main land mass is highly compact. In P. Haggett's book, *Locational Analysis in Human*

Figure 144

Geography, published in 1965, reference is made to Gibb's formula $4A/\pi l^2$, where A is the area and l the length of the longest axis. This is very similar to Cole's definition and, indeed, must yield identical results in those cases where the circumscribing circle is tangent to the area at the extremities of the longest axis, though in all other cases it will give greater values.

Blair and Bliss propose the formula:

$$C = \frac{\text{area of region } R}{\sqrt{2\pi \int_R r^2 \, dR}}$$

where r is measured from the centroid of R. For a straight line C is zero, and for the circle of radius r we would have:

$$\frac{\pi r^2}{\sqrt{2\pi \cdot \pi r^2 \cdot \dfrac{r^2}{2}}} = 1$$

The circle being the most compact figure, witness soap bubbles and the canning industry, this gives us a very convenient scale of measurement from 0 to 1. The indices with the country outlines in Figure 142 have been calculated by this formula.

Mathematical readers will instantly recognize in $\int_R r^2 \, dR$ the second moment of area of R about its mean centre or, alternatively, the moment of inertia of a lamina R of unit surface density. Hence the underlying idea is to think of compactness of shapes of the same area as inversely proportional to their swing radii or radii of gyration. We use the same kind of measure in evaluating the root mean square value of, say, an electrical current over its cycle and also in calculating the standard deviation of a set of statistical data.

In the same way we could borrow from statistics the idea of the mean deviation, and define the compactness of the region R in terms of the mean dispersion of its area from its centroid. For example, if we were to write

$$C = \frac{\text{perimeter of region } R}{3\pi \cdot \dfrac{1}{\text{area of } R} \cdot \int_R r \, dR}$$

we should have, for the circle,

$$\frac{2\pi r}{3\pi . \dfrac{1}{\pi r^2} \displaystyle\int_0^r x . 2\pi x \, dx} = 1$$

and, for extremely attenuated areas, values tending to zero. However, as with Pounds' and Millar's methods, this method depends too critically upon the nature of the boundary.

The various measures of compactness do, of course, give different values because we are using quite different units of measurement. A comparison of these indices for the circle, square, and equilateral triangle is interesting.

Indices of compactness

	Pounds		Millar	Cole	Blair and Bliss
	boundary length / area (scaled to give 1 for circle)		$\dfrac{area}{\begin{array}{c}area\ of\\circle\\of\ same\\perimeter\end{array}}$	$\dfrac{area}{\begin{array}{c}area\ of\\smallest\\circumcircle\end{array}}$	$\dfrac{area\ R}{\sqrt{2\pi \int_R r^2 \, dR}}$
	1		1	1	1
	1.13 (area const.)	or 1.27 (perimeter const.)	0.79	0.64	0.98
	1.29 (area const.)	or 1.65 (perimeter const.)	0.60	0.41	0.91

Millar's and Cole's methods give good discrimination between shapes which have obvious differences in compactness. Blair and Bliss's, while not suffering the disadvantages of the others in unusual cases, nevertheless tends only to yield slight discrimination between quite different shapes. This could be improved by squaring their measure.

The key question, as always, is 'Why bother?'. Yet ever since D'Arcy Thompson published his celebrated *On Growth and Form*, in which he accounted for the sizes and shapes of various species, it has clearly been accepted as legitimate to seek a reason for shape. Why are the parishes in Cheshire compact, while nearly all those in Wiltshire and Lincolnshire are elongated? What is the effect of time on the shape of a political unit? Does it tend to become more or less compact? When lines of communication have to be as short as possible, the greater the compactness the better. But, in these days of supersonic air transport and radio satellites, is distance a major factor (except in exceptional cases such as Pakistan)? The compactness indices for Malaysia and the European Free Trade Area are both small; amalgamation has occurred nevertheless. But communications are not the only factor involved.

Geomorphologists are investigating the effect of physical and geological factors on shape. There is undoubtedly a relationship between shape and relief, scarp slopes tending to produce elongated units of division. Then again there are the problems of security and defence, which are almost directly proportional to the length of frontier per acre of state. The whole business of administration of a country or county or rural district must depend in some way on shape. It would perhaps be enlightening to investigate the compactness of the regional divisions proposed by the Maud Commission and to compare these with the county areas. Indeed, is there a tendency for the whole to be more compact than the parts—the country than the counties, the dioceses than the parishes? Does the compact country have more neighbours than others, and are inland countries on the whole more compact than those partly bounded by oceans? (Figure 142 shows that Chile and Norway are both half surrounded by sea.) What is the correlation between the compactness of a city and its sphere of influence, an urban area and its rural district? Is there any significant correlation between population density and compact shape? Answers to all these questions depend upon being able to measure compactness in the first place.

Factor analysis

We have seen how, for example, the populations of various regions may be represented by a topologically distorted map. Of course, we could simply use a scatter of dots on an ordinary topographical map; the densest population would correspond to the closest or heaviest dots. Suppose, however, that we wish to look at the interacting effects of several variables. We might, for instance, have figures relating to temperature, rainfall, relief, population, occupations, compactness, number of telephones per household, number of cars per household, infant mortality, and so on, for all the states in the US. For each statistic we could draw a demographic map and then compare them two by two. But even in our short list above of only nine variables, there are $^9C_2$, or 36 different comparisons to be made—a rather daunting task!

The analysis of whole sets of factors is now an important concern of geographers, and the means by which they determine subsets of variables which give rise to similar types of correlation is called *factor analysis*. The technique demands a computer, and the determination and significance of the eigen values of the consequent matrix of correlation coefficients is beyond the scope of this chapter. However, we shall look at an extremely simple factor analysis of the north midland counties. The factors to be considered are: (A) the 1961 figures for the population density in persons per acre, (B) the percentage of the population who were born in the country, (C) the average domestic rateable value, and (D) the infant mortality per thousand.

	A	B	C	D
Notts.	1.7	68	53	22
Derby.	1.4	71.5	50	21
Leics.	1.3	70	63	20
Northants.	0.7	67	57	20
Lindsey	0.5	72.2	52	23
Holland	0.4	78	45	20
Kesteven	0.3	64	45	19
Rutland	0.2	42	44	16

Table 1

We now calculate the rank coefficient of correlation between all possible pairings of the factors A, B, C, and D. To do this we tabulate not the values in Table 1, but their *rank*, or order of size. Since the figures for A are already in descending order we simply rank them from 1 to 8.

		A	B	C	D
	Notts.	1	5	3	2
	Derby.	2	3	5	3
Table 2	Leics.	3	4	1	5
	Northants.	4	6	2	5
	Lindsey	5	2	4	1
	Holland	6	1	$6\frac{1}{2}$	5
	Kesteven	7	7	$6\frac{1}{2}$	7
	Rutland	8	8	8	8

The coefficient of rank correlation r between N pairs of values is given by:

$$r = 1 - \frac{6\,\Sigma\,d^2}{N(N^2-1)}$$

where the values of d are the differences in rank between each pair. Table 3 shows the computation of all pairwise correlations between A, B, C, and D. (We should note that, since in each case $N = 8$, the formula reduces to

$$(r = 1 - \frac{6\,\Sigma\,d^2}{8 \times 63} = 1 - \frac{\Sigma\,d^2}{84})$$

	A & B	A & C	A & D	B & C	B & D	C & D
	d^2	d^2	d^2	d^2	d^2	d^2
	16	4	1	4	9	1
	1	9	1	4	0	4
	1	4	4	9	1	16
	4	4	1	16	1	9
	9	1	16	4	1	9
	25	$\frac{1}{4}$	1	30.25	16	2.25
	0	$\frac{1}{4}$	0	0.25	0	0.25
	0	0	0	0	0	0
$\Sigma\,d^2 =$	56	22.5	24	67.5	28	41.5
$\dfrac{\Sigma\,d^2}{84} =$	0.67	0.27	0.29	0.80	0.33	0.49
$1 - \dfrac{\Sigma\,d^2}{84} = r =$	0.33	0.73	0.71	0.20	0.67	0.51

Table 3

Finally, in Table 4, we have the matrix of correlation coefficients (here expressed as percentages)—a very useful condensation of the inter-relationships between the factors involved, since it expresses succinctly, in one set of figures, all possible pairwise relationships that arise by comparing four maps (each illustrating one factor) two at a time.

	A	B	C	D
A	100	33	73	71
B	33	100	20	67
C	73	20	100	51
D	71	67	51	100

Table 4

Of course, a high correlation does not necessarily indicate a causal relationship between two factors; the correlation between teachers' salaries and road deaths, which have both risen steadily over the last ten years, would be high, but no one would suggest that the former causes the latter! The point is, that if a causal connection can be proved by independent scientific means, then the coefficient of correlation provides a convenient mode of measurement. However, where a high correlation coefficient exists, it is sometimes worth investigating whether this signifies more than coincidental trends.

This is really the attitude of the geographer using factor analysis. The patterns revealed by the matrix of correlation coefficients some-times indicate unexpected relationships between factors such as population density, mortality rates, rainfall, mean income, and so on. More important, where a whole range of variables is under considera-tion, then certain subsets of these may give rise to similar types of correlation. In fact scattergraphs of these show up as clusters of points along one of the eigenvectors of the matrix. Geographers normally regard a correlation coefficient of 0.3 (30%) or more as indicative of some connection; a coefficient of 0.5 (50%) would be regarded as a fairly positive pointer. In the highly simplified example considered here, the correlation between population density A and domestic rateable value C is very high (73%)—as we might well expect. Nearly as strong, though less expectedly, is the connection between popula-tion density A and infant mortality D (71%). While it might be expected that the healthiest infants are born to those who live in the wide open spaces, medical facilities are undoubtedly more accessible in urban areas. Even more difficult to understand is the high correlation (67%) between the infant mortality rate D and the percentage, B, of the population born in the country. The smallest correlations are between

A and B and between B and C. Reasons for the first are fairly obvious; highly mobile 'units' of population tend to move between urban areas, while country dwellers tend to live and die in the same place. The second, that is, the low correlation between B and C, is simply a consequence of the first, for if B and A have low correlation (33%) and A and C have high correlation (73%), then B and C may obviously be expected to have low correlation.

The scattergraph of A and C shown below indicates a clustering about the broken regression line—a similar situation to that of the A/D scattergraph. Likewise, scattergraphs of B against A and C would be similar. Even with this simple example the correlation matrix clearly

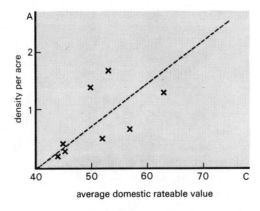

Figure 145 A scatter graph of A and C

abstracts and highlights all kinds of relationships which it might be fruitful geographically, or even socially, to pursue. Factor analysis of dozens of variables over, say, a whole country or continent can clearly produce answers to old questions and open up entirely new lines of enquiry not only for geographers but for geologists, anthropologists, and many others.

Areal point patterns

People do not distribute themselves uniformly over the surface of a country or county. Being gregarious they live together in villages and towns. Industrial economics has indeed forced them in some areas to form larger units in the great urban conurbations. The way in which towns or communities are distributed is interesting and often depends on significant geographical features—rivers, terrain, proximity to the sea, and so on. Sometimes the distribution is very even, like holes in a peg board; sometimes it is very clustered. Between these two extremes the distribution may be quite random or nearly so. Nowadays geographers need to measure the distribution of town locations or point patterns and they do so by what is called the 'nearest neighbour' technique, a method developed by P. J. Clark and F. G. Evans in 1954, for ecological studies, and by L. J. King in a study of urban settlement spacing in the US.

Suppose a is the area of a country or county with N towns. We define $\bar{r}_A$ as equal to $\Sigma r/N$ where Σr is the sum of the shortest distances between each town and its nearest neighbour; so $\bar{r}_A$ is the average of all these shortest distances Now it can be shown that for a random distribution $\bar{r}_E = \sqrt{(a/N)}$. We now work out the 'nearest neighbour statistic' $R = \bar{r}_A/\bar{r}_E$. This value R ranges in practice between 0 and 2.15. A value below 1 shows a distribution more clustered than random, a value of 1 shows random placing, while a value greater than 1 indicates greater regularity or evenness of spacing than when random.

For Brazil the calculations shown below indicate towns more clustered than random—in this case, round the coast. India has roughly the same 'nearest neighbour statistic', 0.6, while the value 1.07 in East Sussex, for instance, reflects a scatter of towns slightly more regular than random.

distances between Brazilian towns

Manaus	7	Fortaleza	5	Belo Horizonte	2	Rio de Janeiro	1
Porto Velho	7	Recife	5	São Paulo	1	Curitiba	3
Belem	3	Salvador	6	Santos	1	Porto Alegre	3
São Luis	3	Brasilia	5	Niteroi	1	Rio Grande	3

calculations

Area $a \simeq 570$

Number of places $N = 16$

$$\bar{r}_E = \frac{1}{\sqrt{\dfrac{N}{a}}} \simeq 6$$

$$\bar{r}_A = \frac{\Sigma r}{N} = \frac{56}{16}$$

$$R = \frac{\bar{r}_A}{\bar{r}_E} = \frac{7}{12} \quad \text{or} \quad 0.6 < 1$$

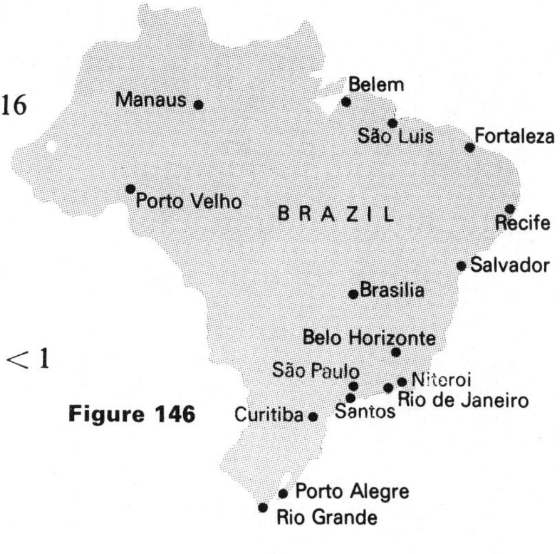

Figure 146

Hence these towns in Brazil are more clustered than randomly distributed.

distances between Indian towns

Amritsar	2	Patna	8	Diu	3	Goa	6
Chandigarh	2	Tezpur	2	Daman	2	Bangalore	3
Bikaner	4	Ledo	3	Bombay	2	Mysore	2
Delhi	3	Kohima	3	Nagpur	5	Mahé	2
Jaipur	3	Shillong	2	Cuttack	1	Madras	2
Jodhpur	4	Imphai	3	Bhubaneswar	1	Pondicherry	2
Lucknow	7	Calcutta	6	Poona	2	Trivandrum	7
		Bhopal	5	Hyderabad	8		

calculations

Area $a \backsimeq 900$

Number of places $N = 30$

$$\bar{r}_E = \frac{1}{\sqrt{\dfrac{N}{a}}} = \sqrt{30} = 5.5$$

$$\bar{r}_A = \frac{\Sigma r}{N} = \frac{105}{30} = 3.5$$

$$R = \frac{\bar{r}_A}{\bar{r}_E} = 0.6$$

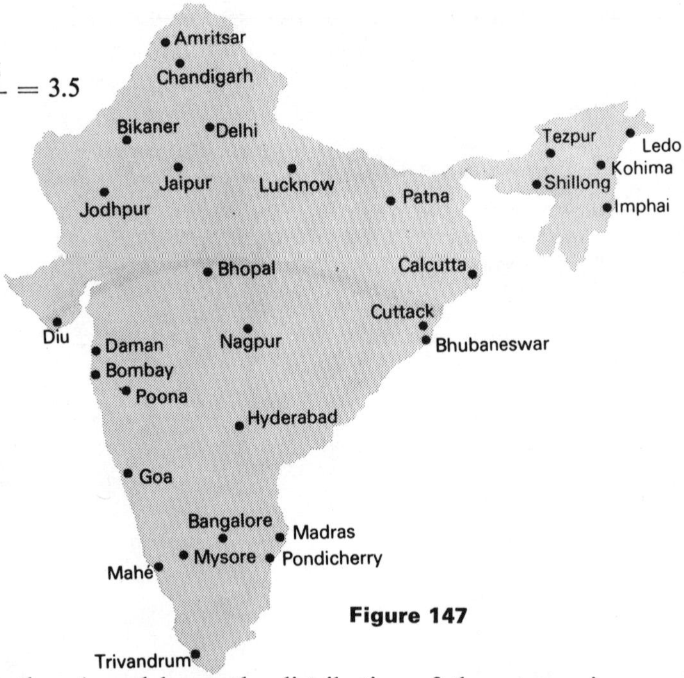

Figure 147

Again R is less than 1, and hence the distribution of these towns in India is more clustered than random.

CHAPTER NINE

Population—7 000 000 000 in A.D. 2000!

'It has taken just one decade from Hiroshima for the world to face up resolutely to the implications of atomic war. Can we hope that it will take no more than a decade from the 1954 World Population Conference in Rome for the world to face up equally resolutely to the implications of world population?'   JULIAN HUXLEY, 1956

Well, can we? And *must* it be a bang or a starved whimper?

World population

One of the greatest problems we face is how to plan for the rapidly increasing world population. It is, of course, no new problem. As long ago as the birth of Christ, governments were aware of the need to keep a check on numbers, and, indeed, at that first Christmas a census was in progress in Palestine. Records of this kind were based solely on counting and there is little evidence to suggest any attempt to forecast the future growth of population.

It was not until 1798, when Thomas Robert Malthus published his historical tract *An Essay on the Principle of Population as it Affects the Future Improvement of Society*, that there was any kind of attempt to predict mathematically the growth of population. This essay argued that as the population increases in geometrical progression and food supplies increase in arithmetical progression, man must eventually outstrip his sources of sustenance. Inevitably, Malthus said, 'population growth must be limited by checks of vice and misery'. The essay aroused bitter controversy; it offended those who believed in the ultimate perfectibility of man, and irritated those who had responsibility for the welfare of the growing number of victims of the Industrial Revolution. Malthus was a nuisance—'a remembrancer of unpleasant facts'.

Later attempts to predict population growth have been based on more valid and more sophisticated mathematical models. The current statistics of the Population Reference Bureau of the United States are probably as reliable as any. We see the magnitude of the problem and the formidable proportions it will assume by the year 2000. All we have to do now is to solve it! The table below is a self-explanatory extract from the 1968 statistics of the Bureau.

	popula- *tion* *1968*	*millions* *population* *projections* *for 2000·* *A.D. at* *present* *growth* *rates*	*% rate* *of* *annual* *increase*	*no. of* *years to* *double* *popula-* *tion*	*average* *life* *ex-* *pectancy*	*% of* *illiter-* *ates in* *popula-* *tion* *over 15* *years*	*per* *capita* *income*
World	3479	6577	2.0	35	53	39	493
Africa	333	689	2.3	31	43	82	123
Asia	1943	3925	2.2	32	50	54	128
North America	222	315	1.1	63	71	2	2793
Latin America	268	690	3.0	24	60	34	344
Europe	455	569	0.7	100	70	5	1069
Oceania	18.5	34	1.8	39	71	12	1636
Russia	239	338	1.1	63	70	0.2	928
(UK)	55.8		0.5	140	71		1451

Some individual countries showing extreme values of these statistics are worth noting.

Kuwait	0.5		5.1 (max.)	14			3184 (max.)
East Germany	17.1		0.2 (min.)	350	71		1240
Netherlands					74 (max)		
Guinea ⎰West					25–35 ⎫		
Mali ⎱Africa					30–35 ⎬		
Togo					30–40 ⎭		
					(min.)		
Malawi							38 (min.)

Figure 148 is a population bar chart showing 1968 population and additional population by A.D. 2000 for the world regions included in the table.

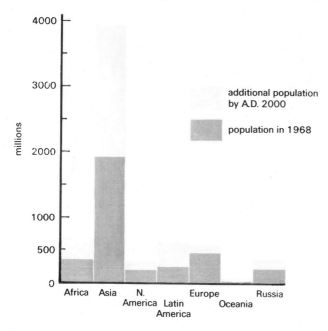

Figure 148 Population bar chart

In 1850 the world's population was estimated at 1000 million; by 1925 it had doubled. In Australasia and the Western Hemisphere the populations have steadily increased since 1900, but those of China and the rest of Asia (excluding Russia and Japan) have shown a far greater rate of expansion. In many cases this is due to improved agriculture and food production and to the spread of modern medical knowledge. The archaic notion 'nature abhors a vacuum' does not apply to population growth; existing population density does not seem to inhibit population growth until a very high value indeed is reached. Thus we find high population density but only moderate growth in Japan and Europe (excluding Russia); in the Caribbean, Central, South, South-East and East Asia we have high density and rapid growth; while in North America, southern South America, Australasia, and Russia the low density and moderate growth contrasts with low density but rapid growth in Africa, Central America, the Middle East, and the Pacific islands.

A table giving some of the current and widely ranging population densities is shown below.

	persons per square mile
Netherlands	890
England	849
Belgium	772
Japan	650
South Korea	595
UK	560
Canada	4.8
Australia	4.75

Figure 149 is an agricultural production/population graph showing percentages of the world's total.

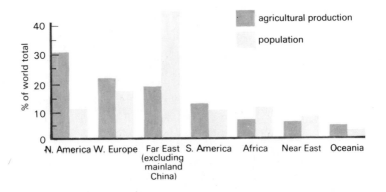

Figure 149 Population and agricultural production bar chart

Figure 150 is a nomograph for calculating population density. Given the population A and the area in acres B, then AB cuts the central scale at P, the population density per square mile. To take another example, for C and D, the population density per square mile is:

$$\frac{20}{3200 \div 640} = \frac{20}{5} = 4 \text{ persons per square mile}$$

This is given directly by Q.

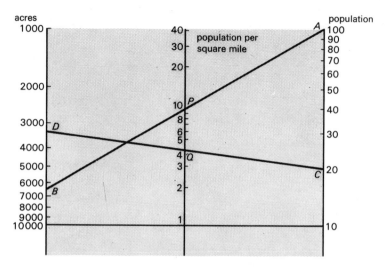

Figure 150 Nomogram for calculating population density

As we saw for the Ridings of Yorkshire in the section *Lorenz curves* in Chapter 8, there has been a tendency towards less even distribution of population. Whether accounted for by the gregariousness of humans, the opportunities for work, or simply the gravitational pull of dense masses, heavily populated areas have tended to become even more so. Towns have grown; rural areas have become more sparsely peopled. Where other external factors have reinforced such tendencies, as did, for example, the partition of India in 1947 with its effect upon the population of Calcutta, overcrowding has sometimes reached nightmare proportions.

Population movements

For various reasons, significant movements of population have occurred in recent centuries. The New World represented for many new hope and opportunity, freedom from persecution and hunger. To America in the seventeenth century emigrated nearly half a million people from England. Led by the Pilgrim Fathers, who set sail in 1620, they settled mainly in New England, Virginia, and Maryland. During the eighteenth century a further 1½ million Irish and Scots crossed the Atlantic, while in the nineteenth century the influx was swollen by immigrants from Germany, Austro-Hungary, and Italy. Indeed, between 1900 and 1920 it is estimated that some 14 million Europeans moved to the US, others going to Canada and South America. During the seventeenth century the French started to settle in Quebec and this tendency for French immigration has continued. The warmer latitudes attracted the southern Europeans; during the later nineteenth and twentieth centuries $4\frac{1}{2}$ million Spanish, Portuguese, and Italians migrated to Argentina and Brazil.

From the sixteenth century to the early years of the nineteenth, the slave trade flourished. During this time at least 20 million West Africans were forcibly deported, some to the tobacco, sugar, and coffee plantations of Brazil and the Caribbean islands, others to the cotton fields of the southern states of America.

Numerous other small migrations have occurred. In the sixteenth century the French started to settle in Algeria, while in 1652 Dutch settlers moved into South Africa. The gold rush of 1850 to 1860 saw many Europeans emigrating to Australia and New Zealand; diamond deposits discovered in South Africa in 1880 precipitated Cecil Rhodes's trek to Salisbury in 1889 and the settling of Europeans in Central and East Africa.

Migrations of population have also affected Asia. During the nineteenth century millions of Chinese settled in Malaysia, Indonesia, Burma, and Ceylon, while some Indians emigrated to Malaysia and Africa. The greatest population movement of recent years occurred when the state of Israel was formed in 1948; it is estimated that during 1949 240 000 Jews returned to the Holy Land.

Apart from these intercontinental and transcontinental migrations of population there have also occurred major population movements within certain countries. Apart from the universal drift to the cities, major population transfers have occurred in Russia, India and the United States. In Russia successive five-year plans have virtually directed people to Kazakhstan and to the east of the Ural Mountains. In India, during the two years from 1947 to 1949, following the partition of the British Indian Empire into India and Pakistan, at least 10 million Hindus and Moslems moved from one state to the other. In America the cry 'Go West, young man' reflected a mass movement of white population into California and Florida; this was accompanied by a steady northward drift of negroes from the southern states.

Population checks

As we have seen, population changes have occurred widely through migration and this phenomenon has many causes. Other factors too, such as disease, famine, war, and natural catastrophes have produced population checks and changes. Man's oldest enemy, plague, has struck frequently—against the Philistines in the eleventh and fourth centuries B.C., in the Emperor Justinian's reign in A.D. 542 and, most catastrophically of all, between 1348 and 1351 when the Black Death took a toll of 20 million out of 85 million lives in Europe. In fact by 1400 one-third of the population had died and the loss was not made good for two centuries! We associate cholera with the East, but between 1831 and 1866 200 000 people died of it in Britain alone. As recently as 1953 800 000 people in India died of malaria. However, none of these can compare with the influenza epidemic of 1918 which, it is estimated, took 100 million lives throughout the world.

The worst famines on record have been in India in 1769 when 10 million starved to death, and in 1877 in China when 9 million died. A famine nearer home, and one which, in proportional terms, was far worse, was the Irish potato famine of 1846, which left $2\frac{1}{2}$ million dead.

Other checks to population may be laid at man's own door: the carnage of the Third Punic War of 146/9 B.C., the systematic genocide of $18\frac{1}{2}$ million Mexicans by the Spanish conquistadors between 1519 and 1548, the First World War with its 8 million dead, the Second with its 22 million losses—including the bombing of Hiroshima (80 000 died).

By contrast, in the famous calamity, the Great Fire of London, it is said that only 6 people died.

The population of England

The earliest record of population we have is contained in the Domesday Book, William the Conqueror's survey of 1086. In this, only adult males who were heads of households were counted: monks, soldiers, and tradesmen were ignored. However, extrapolation from this basis suggests that the population of England at that time was between $1\frac{1}{2}$ and 2 million. Certainly there is clear evidence of its distribution. The most densely populated areas were, in descending order, East Anglia, Kent, and North Lincolnshire. Norfolk and Suffolk were particularly heavily peopled, though the Fens and Weald were uninhabited.

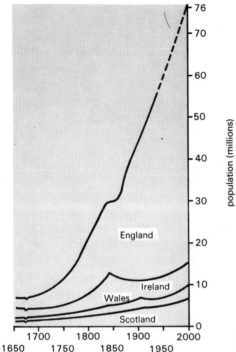

Figure 151 Population of Great Britain and Ireland

Figure 151 shows the beginnings of our 'population explosion' in Great Britain and Ireland in 1650. Up to this time, throughout the Dark Ages, birth and death rates, both high, had roughly balanced; the population remained small and was, apart from in the eastern counties, sparsely spread across the land. Indeed, at the time of the Black Death in the fourteenth century, over 25% of the population was wiped out and it took the country more than a hundred years to make good the loss. In the year of the Plague—which hit London in 1664/5 and subsequently spread—one-tenth of London's population (of half a million) died.

From the eighteenth century on there was a drift to the towns. Mortality rates were slow to fall but by this time the birth rate had increased. The resulting net percentage rise in the population in the seventeenth and eighteenth centuries increased thereafter even more rapidly, though in Wales and Scotland the population growth rate was not comparable.

There are many reasons for this marked net growth of population in England from 1700 onwards. The following events and developments clearly combined to produce this result, whether by boosting the birth rate or prolonging life expectancy.

1710 Townshend's improved agricultural methods
1720 Growth of hospital foundations
1785 Onset of the Industrial Revolution
1796 Jenner's first vaccination
1802 The first Factory Act
1833 Employment of children under nine years forbidden and the restriction of the working hours of those between nine and eighteen
1848 Chadwick's Public Health Act
1861 Pasteur's work on bacterial infection
1867 Lister's discovery of antiseptics with the consequent dramatic decrease in post-surgical mortality
1875 Compulsory vaccination against smallpox (later repealed)
1882 Isolation of the tubercle bacillus by Koch
1895 Roentgen's discovery of X-rays
1936 Development of sulphonamides, penicillin and antibiotics
1948 Establishment of the National Health Service

Today, half the population survive into their seventies, though of those who do women form a higher proportion.

The distribution of the sexes is also interesting. On parts of the south coast of England, particularly areas to which the elderly retire, the proportion of women to men is occasionally in excess of 130 : 100 for the very reason we have seen above. However, over the whole of Ireland, except in Dublin and Belfast, and in parts of England, men outnumber women. In Rutland, for example, there are 84 women to every 100 men.

In previous sections of this chapter we have looked at world population movement. Internally in Great Britain there has been a strong drift of population from north to south; indeed, between 1951 and 1961 more than a million people from the North and the Midlands moved into Greater London and its surrounds. In these influx areas birth rates have increased; in the afflux areas mortality rates have increased. On

Figure 152 Birth-rate variations over the period 1871–1971

balance, as the graph in Figure 152 shows, there is some tendency for the mean birth rate to rise.

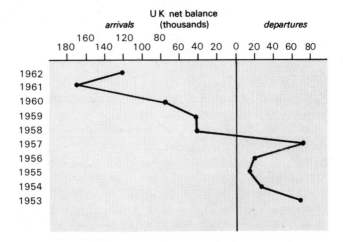

Figure 153 UK population movements over the period 1953–62

External population movements are shown in Figure 153. Between 1953 and 1957 there was a net loss of population through emigration to Australia, Canada, New Zealand, Rhodesia, South Africa, the West Indies, and other Commonwealth and foreign countries. Since then, however, the flow of immigrants from India, Pakistan, and Central Europe has increased to such an extent that it has become necessary to impose some restriction on immigration.

The Malthusian nightmare

As we have seen, the first essay on population growth and its implications, published in 1798, brought notoriety rather than acclamation to its author, the Reverend Thomas Robert Malthus (1766–1834). After travelling in Germany, Sweden, Norway, Finland, and Russia, Malthus produced a second edition of the essay in 1803, in which he took the view that checks to population were not to be regarded as 'insuperable obstacles'. This was no new line of thought. Condorcet in France, writing at about the same time, sanguinely anticipated the practice of birth control. What was new was Malthus' introduction of the speculative question of moral restraint. He asked: What checks on population are operating? He classified them as 'positive checks' and 'preventive checks'—the latter included 'improper acts to conceal the consequences of irregular connections'. This last presumably referring to abortion. Malthus even discusses contraception as 'something else as unnatural as promiscuous concubinage'.

In his first essay, Malthus investigated the impediments in the way of mankind's progress towards happiness. His central thesis was 'the constant tendency of all animated life to increase beyond the nourishment prepared for it'. He writes:

'The population of the Island is computed to be about seven millions, and we will suppose the present produce equal to the support of such a number. In the first twenty-five years the population would be fourteen millions, and the food being also doubled, the means of subsistence would be equal to this increase. In the next twenty-five years the population would be twenty-eight millions, and the means of subsistence only equal to the support of twenty-one millions. In the next period, the population would be fifty-six millions, and the means of subsistence just sufficient for half that number. And at the conclusion of the first century the population would be one hundred and twelve millions and the means of subsistence only equal to the support of thirty-five millions, which would leave a population of seventy-seven millions totally unprovided for.'

The conclusion to which this leads is, of course, the necessity to check the population increase by artificial means. Malthus continues:

'The effects of this check (i.e. the constant operation of the strong law of necessity) on man are more complicated. Impelled to the increase of his species by an equally powerful instinct, reason interrupts his career and asks him whether he may not bring beings into the world for whom he cannot provide the means of subsistence. In a state of equality, this would be the simple question. In the present state of society other considerations occur. Will he not lower his rank in life? Will he not subject himself to greater difficulties than he at present feels? Will he not be obliged to labour harder? and if he has a large family, will his utmost exertions enable him to support them? May he not see his off-spring in rags and misery, and clamouring for bread that he cannot give them? And may he not be reduced to the grating necessity of forfeiting his independence, and of being obliged to the sparing hand of charity for support?

'These considerations are calculated to prevent, and certainly do prevent, a very great number in all civilized nations from pursuing the dictate of nature in an early attachment to one woman. And this restraint almost necessarily, though not absolutely so, produces vice.'

Discussing the same problem nearly 200 years later, Desmond Morris, in his famous book *The Naked Ape*, published in 1967, writes:

'If certain sexual patterns interfere with reproductive success, then they can generally be referred to as biologically unsound. Such groups as monks, nuns, long-term spinsters and bachelors and permanent homo-sexuals are all, in a reproductive sense, aberrant. Society has bred them, but they have failed to return the compliment.

'This biological morality ceases to apply under conditions of popula-tion overcrowding. When this occurs the rules become reversed. We know from studies of other species in experimentally overcrowded conditions that there comes a moment when the increasing population density reaches such a pitch that it destroys the whole social structure. The animals develop diseases, they kill their young, they fight viciously and they mutilate themselves. No behaviour sequence can run through properly. Everything is fragmented. Eventually there are so many deaths that the population is cut back to a lower density and can start to breed again, but not before there has been a catastrophic upheaval. If, in such a situation, some controlled anti-reproductive device could have been introduced into the population when the first signs of over-crowding were apparent, then the chaos could have been averted . . . Our own species is rapidly heading towards just such a situation. We have arrived at a point where we can no longer be complacent. The solution is obvious, namely to reduce the breeding rate without inter-

fering with the existing social structure . . . The need to reduce drastically the reproduction rate now removes any biological criticism of the
non-breeding categories such as monks and nuns, the long-term
spinsters and bachelors and the permanent homosexuals . . . Provided
they are well-adjusted and valuable members of society outside the
reproductive sphere, they must now be considered as valuable non-
contributors to the population explosion!'

Elsewhere, writing of the northern states of the US, Malthus com-
ments on the exceptionally high birth rate to be found in certain areas
and refers to the great Leonhard Euler's passing interest in the mathematics of the problem. In the northern part of the US

'the population has been found to double itself, for above a century
and a half successively, in less than 25 years. Yet even during these
periods, in some of the towns, the deaths exceeded the births . . . In the
back settlements . . . the population has been known to double itself
in 15 years . . . According to Euler, assuming mortality of 1 : 36, if
births : deaths are 3 : 1, and the period of doubling is 12 years . . . Sir
William Petty supposes a doubling possible in so short a time as 10
years.'

The checking of Euler's calculations is a simple matter. We assume,
as stated, an annual mortality rate of 1 in 36 and a ratio of births to
deaths of 3 to 1; then if there are p people to begin with, the number
after the first year is:

$$p - \frac{p}{36} + 3 \times \frac{p}{36} = \frac{19}{18}p$$

After 2 years there are $\left(\frac{19}{18}\right)^2 p$, and so on. Hence, after n years there are

$\left(\frac{19}{18}\right)^n .p$ people. If, in n years, the population has doubled, we write:

$$\left(\frac{19}{18}\right)^n .p = 2p$$

or

$$n \log \frac{19}{18} = \log 2$$

So we can calculate:

$$n = \log 2/\log \frac{19}{18} = \frac{0.3010}{0.0235} = 12.80$$
$$\simeq 12\tfrac{4}{5} \text{ years}$$

This was Euler's result.

Population regulation

In this section we shall look at one or two simple mathematical models of population growth under various conditions. It is obvious to us now that Malthus's model was greatly over-simplified. His view that, like amoebae in a limitless tank under ideal conditions, humans increase in geometrical progression cannot be sustained in the face of historical fact. The population growth rate is now tending to rise all over the world. In some places this is because of material prosperity and social security; in most places it is simply because modern medicine and improved methods of food cultivation have lengthened the life span and clipped the mortality rate. But there have been spasmodic alterations in the population growth rate; to such natural and man-made catastrophes as the plague, the influenza epidemic of 1918, and the two World Wars, humankind has reacted instinctively, to preserve itself and replace the loss. Such events have nearly always been followed by a rise in the birth rate. On the other hand, conditions of famine and gross overcrowding are often accompanied by a fall in the rate, whether naturally or deliberately contrived. India, aware of its own great population problems, has in recent years set up many family planning centres and sterilization clinics. Consequently, in certain parts of the subcontinent, the birth rate has fallen from 4 to 2.5 per thousand.

The supply of food is clearly a factor likely to influence the growth rate of population. Indeed, for some species there is a significant positive correlation between the food they consume and the offspring they produce. But the situation is often more complicated, for in the selfsame world we have, co-existing, many populations of different species. Some are preyed upon, and others prey—the most efficient and ruthless predator of all being man. Where there are problems of space, food, and enemies, the growth rate of a population can hardly remain constant.

Let us look at some simple mathematical models showing various cases. We shall assume first a population with a single summer breeding season for the adults, who die before the next summer. Let us suppose that, on average, x_n females each produce r female babies, who all survive until the following summer. Then:

$$x_{n+1} = rx_n$$

where n records the generation number.

And the growth of female population in one year is:

$$x_{n+1} - x_n = (r - 1)x_n$$

that is,

$$\Delta x_n = (r - 1)x_n$$

(where Δ denotes the *difference*—in this case between successive values of x at generations n and $n + 1$).

If r is constant we have:

$$x_n = rx_{n-1}$$
$$x_{n-1} = rx_{n-2}$$
$$.$$
$$.$$
$$.$$
$$x_2 = rx_1$$

and so

$$x_n = r^{n-1}.x_1$$

If we say that $x_1 = 2$, $r = 3$, and $n = 6$, we should have:

$$x_6 = 3^5.2$$
$$= 486$$

Clearly, when $r > 1$, the population of females increases without limit; when $r = 1$, $x_n = x_1$, and the population remains constant, while if $r < 1$ the population will gradually disappear completely.

Generally, however, r will depend on x_n. When x_n is very small and the population very sparse, and perhaps highly scattered, some of the females may well be unable to find mates. On an average, r may be very small. Again, when x_n is very large, overcrowding and under-nourishment will probably set in with a consequent decrease in r. Between these extremes r may increase to some optimum value.

Let us consider the population values $x = a$ and $x = b$, when $r = 1$ and the population is only just replacing itself 1 : 1. The lower value is unstable, for the smaller r becomes the less x_n is the next year, and so on; but at the upper extreme, where overcrowding is occurring, a fall

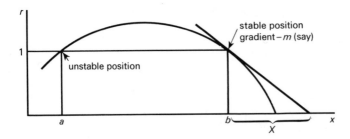

Figure 154

in r will tend to correct the situation with consequent improvement thereafter. Such a situation is stable. The position may be pictured as in Figure 154. Let us look at this upper, stable value of $x = b$ more closely. If we take $x = b$ as a new r-axis and measure population increase X from it, then the equation connecting r and X is $r = 1 - mX$, where $-m$ is the gradient of the tangent at $x = b$. Now we know that the values of x (and X) are really discrete, being annual population values, so we do not really have a continuous curve. Then, if $x = b$, after n generations we should really write:

$$r = 1 - mX_n$$

Also, instead of $x = b + X$, we should really write:

$$x_n = b + X_n$$
$$x_{n+1} = b + X_{n+1}$$

and so on. But:

$$x_{n+1} = rx_n$$

and hence

$$b + X_{n+1} = (1 - mX_n)(b + X_n)$$

or

$$b + X_{n+1} = b + X_n - mbX_n - m(X_n)^2$$

Now since we are really interested in what is happening at $x = b$, the values of X are necessarily small and we may ignore $(X_n)^2$, so the last equation becomes:

$$X_{n+1} = X_n(1 - mb)$$

or

$$X_{n+1} = (1 - C)X_n$$

where mb is replaced by the constant C.

At this generation the increase in population is calculated as follows:

$$\Delta X_n = X_{n+1} - X_n = -CX_n$$

From this, four general cases arise:

(1) If $C < 0$, $\Delta X_n > 0$, and successive departures from the equilibrium position increase without limit. For example, with $C = -2$, $X_{n+1} = 3X_n$ and we have $X_1 = 1$ (say), $X_2 = 3$, $X_3 = 9$, $X_4 = 27$, and so on.
(2) If $0 < C < 1$ equilibrium is approached without oscillation. For example, with $C = \frac{1}{2}$, $X_{n+1} = \frac{1}{2}X_n$ and we have $X_1 = 1$ (unit of population, say), $X_2 = \frac{1}{2}$, $X_3 = \frac{1}{4}$, $X_4 = \frac{1}{8}$, and so on.
(3) If $1 < C < 2$ equilibrium is approached with oscillations of decreasing amplitude. For example, with $C = 1\frac{1}{2}$, $X_{n+1} = -\frac{1}{2}X_n$ and we have $X_1 = 1$ (say), $X_2 = -\frac{1}{2}$, $X_3 = \frac{1}{4}$, $X_4 = -\frac{1}{8}$, and so on.
(4) If $C > 2$, we have oscillations of increasing amplitude. Thus with $C = 4$, $X_{n+1} = -3X_n$, and with $X_1 = 1$ (say), we have $X_2 = -3$, $X_3 = 9$, $X_4 = -27$, and so on.

And then three special cases:

$C = 0$ when $X_{n+1} = X_n$ and the population is steady.

$C = 1$ when $X_{n+1} = 0$ and the population falls to zero never to rise again.

$C = 2$ when $X_{n+1} = -X_n$ and the population oscillates with constant unit amplitude of $+1$ and -1.

Logarithmic growth

To populations where birth and death take place continually, the previous model does not apply. Let us now imagine a population of initial size P and, after time t, of x people, in which the chance that any given individual gives rise to another during a time interval δt is $b\, \delta t$ (b for birth), and the chance of dying (m for mortality) in a time interval δt is $m\, \delta t$. Such assumptions allow for no differentiation in age structures in either P or its subsequent value x. Thus the model can only apply to a very uniform set of people or species of the same age and comparable health prospects. But with this restriction we see that if the population change is δx in time δt, we may write:

$$\delta x = (b - m)\, \delta t . x$$

As $\delta t \rightarrow$ have:

$$\int \frac{dx}{x} = \int (b - m)\, dt$$

and thus

$$\log_e x = (b - m)t + C$$

Now when $x = P$, $t = 0$ and so $C = \log_e P$, then

$$x = P\, e^{(b-m)t}$$

and the growth is purely exponential. If the chances of being born or dying are the same—that is, if $b = m$, as it was in England up to about 1650, x remains fairly constant and equal to P and the population remains the same. However, the effect of increasing b and decreasing m is to create the pattern of exponential growth illustrated in Figure 155.

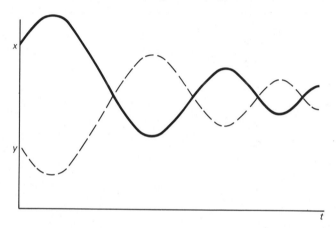

Figure 155

Let us now take into account the limiting factor of food supply and suppose that, as before, we have some initial population P and a food supply sufficient for N people. Clearly we can no longer assume birth and death 'chances' constant. A more reasonable assumption might now be that, at any time t, the rate of increase of population per member of the population x is proportional to $N - x$; that is to say, when the actual population x falls well below N, the birth rate will be high and the death rate low, but as the gap narrows the reverse will be the case. In symbols this becomes:

$$\frac{dx/dt}{x} = k(N - x)$$

where k is the constant of proportionality. Hence:

$$\int \frac{dx}{x(N - x)} = \int k\, dt$$

that is,

$$\int \left(\frac{1}{x} + \frac{1}{N-x}\right) dx = \int kN \, dt$$

that is,

$$\log_e \frac{x}{N-x} = kNt + C$$

Now when $x = P$, $t = 0$, therefore:

$$C = \log_e \frac{P}{N-P}$$

And so:

$$\log_e \left[\frac{x}{N-x} \cdot \frac{N-P}{P}\right] = kNt$$

or

$$\frac{x}{N-x} = \frac{P}{N-P} e^{kNt}$$

or

$$x = \frac{NP e^{kNt}}{N - P(1 - e^{kNt})}$$

Let us look at a special case. We shall suppose that, in some hypothetical country, $N = 10$ million, $x = 8$ million, and the growth rate of population at that time as 4% (or 1/25) per annum. Then, with t in years, we have:

$$\frac{1}{x} \cdot \frac{dx}{dt} = \frac{1}{25} = k(10 - 8) \times 10^6 = k.2 \times 10^6$$

Hence $k = 2 \times 10^{-8}$. What is the situation 10 years later? Now we have $N = 10$ million, $P = 8$ million, $k = 2 \times 10^{-8}$, and $t = 10$, and we want to find x. From our formula, we have:

$$x = \frac{(10 \times 10^6) \times 8 \times 10^6 \times e^2 \times 10^{-8} \times 10 \times 10^6 \times 10}{10 \times 10^6 - 8 \times 10^6[1 - e^2 \times 10^{-8} \times 10 \times 10^6 \times 10]}$$

$$= \frac{80 \times 10^{12} \times e^2}{10^6[10 - 8 + 8 e^2]}$$

$$= \frac{80 e^2}{2 + 8 e^2} \times 10^6$$

$$\simeq \frac{80 \times 7.42}{61.2} \text{ million} \simeq 9.7 \text{ million}$$

Of course, only if we can find a population which behaves similarly to our model is the latter of any value. More often than not our original assumptions are pure conjecture and remain so.

Other logistic curves have been postulated. One of these, investigated by Raymond Pearl in population studies in 1920, was of the form:

$$N = N_0 \frac{1 + b}{1 + b\,e^{-kt}}$$

In this model the ultimate population size is $N_0(1 + b)$, the maximum growth rate occurring when the population is half this size. Another model, by Benjamin Gompertz, takes the form:

$$N = p\,e^{-be-kt}$$

where $N_0 = p\,e^{-b}$.

Predator and prey

As we have seen, populations are limited by food supply and affected by disease and disaster. Interesting situations arise when we have two populations, one of which (the prey) is the other's (the predator's) food supply. Such situations arise frequently in the animal kingdom, and often a species is both predator and prey. The predator supreme, of course, is man, who preys, for a wide variety of purposes, on many lower species.

For the sake of simplicity, however, let us assume a prey, never short of food for themselves, who multiply at a constant rate, and predators who feed on them. Everything now depends on the relation between the predators' feeding and reproduction rates. Until we know this we cannot construct the appropriate mathematical model. Sometimes the relation is almost one of direct proportion. The following sequence of events may then occur. The predators eat the prey, then multiply accordingly. As the number of prey decreases, the predators must eat less and reproduce more slowly. Eventually so few prey remain that the predators themselves begin to die. Gradually the reproduction rate of the prey overtakes its 'disappearance rate' and the prey population recovers. The surviving predators, finding their food supply reappearing, recover, and the whole cycle recurs, unless at some previous stage one of the populations has vanished altogether.

Here is one simple model in which Xn and Yn are the population sizes of prey and predator respectively after n generations.

$$Xn + 1 = \frac{4Xn}{Yn + 2} \qquad Yn + 1 = \frac{Xn\,Yn}{10}$$

Here, as predators Yn increase, $4/Yn + 2$ and hence $Xn + 1/Xn$, the prey growth rate, decrease together. At the same time $Yn + 1/Yn$, the predators' growth rate, equals $Xn/10$ and thus clearly increases linearly with Xn. Starting with $X_1 = 12$, and $Y_1 = 2$, the first twelve generations work out as follows:

n	1	2	3	4	5	6	7	8	9	10	11	12
Xn	12	12	10.9	9.0	7.0	5.8	5.9	7.4	11.0	17.6	27.7	37.0
Yn	20	2.4	2.9	3.1	2.8	2.0	1.1	0.7	0.5	0.6	1.0	2.7

Setting aside the problem of the units involved, and the meaning of '2.4' predators, we have here an example of the case described above.

Vito Volterra constructed a different model in which x prey and y predators interact continuously in a manner symbolized by these differential equations:

$$\frac{dx}{dt} = x(g - ky) \qquad \frac{dy}{dt} = y(-d + kx)$$

Rewriting them as

$$\frac{dx/dt}{x} = g - ky \quad \text{and} \quad \frac{dy/dt}{y} = kx - d$$

we see that the assumptions made are as follows:

(A) The rate of increase of prey per head is a constant minus k times the number of predators. In other words, the prey have a constant natural growth rate g and a depredation loss directly proportional to the number of predators (k kills per predator).
(B) The rate of increase of predators per head is the sum of two factors—a negative constant d and a term which increases directly with the number of prey available for eating.

The equations cannot be solved analytically for time t, though difference equations to which they approximate roughly can be constructed. However, eliminating t, we can obtain a relation between x and y as follows:

$$dt = \frac{dx}{x(g - ky)} = \frac{dy}{y(-d + kx)} \tag{1}$$

$$\Rightarrow \int \left(-\frac{d}{x} + k\right)dx = \int \left(\frac{g}{y} - k\right)dy$$

$$\Rightarrow g \log y + d \log x = k(x + y) - C \tag{2}$$

where C is a constant.

Writing (1) as

$$\frac{dy}{dx} = \frac{y(-d + kx)}{x(g - ky)}$$

we see that at $x = d/k$ the gradient is zero, and at $y = g/k$ it is infinite. Moreover, for

$$\left.\begin{array}{l} x > \dfrac{d}{k}, \quad 0 < y < \dfrac{g}{k} \\[2mm] 0 < x < \dfrac{d}{k}, \quad y > \dfrac{g}{k} \end{array}\right\} \dfrac{dy}{dx} \text{ is } +ve$$

and

$$\left.\begin{array}{l} x > \dfrac{d}{k}, \quad y > \dfrac{g}{k} \\[2mm] 0 < x < \dfrac{d}{k}, \quad 0 < y < \dfrac{g}{k} \end{array}\right\} \dfrac{dy}{dx} \text{ is } -ve.$$

Thus, for suitable initial conditions, the graph of (2) is a closed loop, described anti-clockwise as the parameter t increases (see Figure 156).

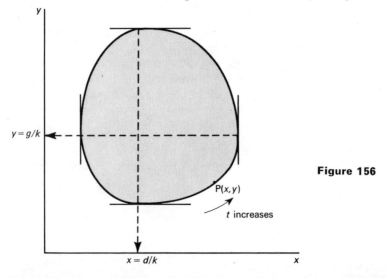

Figure 156

In other words, any point P (x, y) lying on the curve (2) moves round and round the loop without stopping as t increases. It follows that, for given initial conditions, x and y each oscillate indefinitely with constant amplitude as time passes. To draw the graph, given values of g, d, and k and suitable initial values of x and y, is no mean task and involves programming a computer to solve the transcendental equation (2) for, say, values of y over a given range of equispaced values of x.

In his book *Mathematical Ideas in Biology*, J. Maynard Smith examines the more realistic case in which the food supply of the prey is also limited. In the continuous case he shows that the corresponding differential equations are:

$$\frac{dx}{dt} = -\frac{R-1}{r}.x - \frac{CX_E}{r}.y \tag{1}$$

$$\frac{dy}{dt} = \frac{(R-1)(r-1)}{CX_E}.x \tag{2}$$

where $X = X_S + x =$ number of prey, and $Y = Y_S + y =$ number of predators. X_S and Y_S are equilibrium densities of prey and predators, X_E is the equilibrium density of prey in the absence of predators and C is a constant. R and r are the maximum reproductive rates of the species so that:

$$\frac{dX}{dt} = (R - 1)X$$

when X is small and predators are absent, and

$$\frac{dY}{dt} = (r - 1)Y$$

when $X = X_E$. Eliminating t by differentiating equation (1) with respect to t, and substituting from equation (2), we get:

$$\frac{d^2x}{dt^2} + \frac{(R-1)}{r}.\frac{dx}{dt} + \frac{(R-1)(r-1)}{r}.x = 0$$

$x = Ae^{mt}$ is a solution of this equation provided that:

$$m^2 + m.\frac{R-1}{r} + \frac{(R-1)(r-1)}{r} = 0$$

This auxiliary equation has real, equal, or imaginary roots m_1 and m_2, according to whether $\left(\dfrac{R-1}{r}\right)^2$ is greater, equal or less than

$$\frac{4(R-1)(r-1)}{r}$$

that is, as $R >$, $=$, or $< (2r-1)^2$. The solution $x = Ae^{mt} + Be^{mt}$, with m_1 and m_2 real roots, is one of pure damping, x decreasing exponentially with time. In the case where m_1 and m_2 are imaginary roots, the solution takes the form:

$$x = A \exp\left[-\frac{(R-1)t}{2r}\right] \cos(kt + \epsilon)$$

This indicates damped, oscillatory variations in x. The corresponding solution for y is similar, though out of phase with that for x (see Figure 155). Charles Darwin used a similar model to account for the variations in populations of lynxes and rabbits living together and also to explain why there are more bumble-bees near towns.

Age, birthrates and mortality

Our previous examples have taken little account of age structure; we were simply concerned with looking at total population size and the mean birth and death rates. In practice the birth rate varies not only with factors such as food and overcrowding, but also with the age group of the mother. In his famous *Genetical Theory of Natural Selection*, Ronald Fisher quotes Australian population figures for 1911, which suggest that the maximum rate of reproduction for married women occurs at 18, and at 31 it has fallen to half the maximum value. The age of marriage is another crucial factor.

'A bride of 30 may expect but 38% of the family she would have borne had she married at 20, and by 35 the number is further reduced by one-half and is less than 19%. With men the potentiality of fatherhood is usually retained to a considerable age, none the less the age of marriage is still very influential since the most frequent age for brides increases steadily with the age of the bridegroom. For bridegrooms of from 34 to 44 years of age, brides of the most frequent age are very regularly ten years junior to their mates. Using the age of their wives as a basis for calculation, men marrying at 40 to 44 may be expected to have only two-fifths of the number of children of men marrying twenty years earlier.'

Again, at any given age, the fertility of women varies greatly. A. O. Powys, writing in *Biometrika*, quotes family sizes of 10 276 women dying, aged 46 or over, in New South Wales from 1898 to 1902. A bimodal distribution, it includes 1 110 with no children (first mode), 976 with 7 children (second mode), and, at the upper end of the distribution, 3 mothers with 22, 24, and 30 children respectively. Death rates, of course, also vary greatly with age. In fact, life insurance tables are computed on the basis that the death rate at age x, and denoted D_x, is related to the number l_x still living out of the same initial total population l_o, by the equation:

$$D_x = \frac{1}{l_x} \cdot \frac{\mathrm{d}}{\mathrm{d}x}(l_x) = \frac{\mathrm{d}}{\mathrm{d}x}(\log l_x)$$

It is not difficult to see how this is derived. Let us suppose, for instance, that of 100 men, 90 may be expected to survive to 20 and 85 to 30 (see Figure 157(a)). The death rate between 20 and 30, we should say, is $100(90 - 85)/90\%$ per ten years, that is, an average over the ten-year

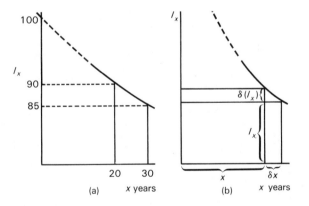

(a) x years   (b) x years   **Figure 157**

period of $100(90 - 85)/(90 \times 10)\%$ per year. We could, indeed, ignore the 100 and give it as a fraction per person per year, as $(90 - 85)/(90 \times 10)$ per person per year. If, instead of age 20, we have age x years, and instead of 90 men, l_x men, and if we think of $\delta(l_x)$ men dying in a period of time δx after age x, as in Figure 157(b) then the formula for the average death rate over the period δx is:

$$\frac{\delta(l_x)}{l_x . \delta x}$$

per person per year. Assuming the death rate to be a continuous function of time, the death rate at any instant in time x years after birth may be expressed approximately by the model

$$\lim_{\delta x \to 0} \left[\frac{1}{l_x} . \frac{\delta(l_x)}{\delta x} \right] = \frac{1}{l_x} . \frac{d(l_x)}{dx} = \frac{d}{dx}(\log l_x) = D_x$$

as quoted above—and with the logarithms taken to the natural base e.

Population forecasts by matrices

Without pursuing these points too far, it is clear that differential birth and death rates may not be ignored if we wish to make any realistic forecast of the size and age distribution in a population. At the same time it is difficult to arrive at a really valid hypothetical relationship between mothers' age groups and population growth rates. Often, the best we can do is to study the pattern of birth and death rates over, say, the last n years and to base our calculations on the assumption that they will remain constant over the next n years, always providing n is not too large.

Population prediction of this kind clearly depends on reliable information. Although ecclesiastical registration records go back to the mid-sixteenth century, they bear no comparison in either accuracy or scope with the demographic information and analyses which have been published in Britain since 1838 by the`Registrar General.

As an example of the same procedure, let us look at the following table. The figures are from an American census but have been greatly simplified to illustrate the method. They show, in thousands to the nearest thousand, numbers of females in three different age groups in a certain State. In the first column we have numbers in these age groups for 1940. The third column shows the number of daughters born to each

age	females in 1940	total females in 1955 (including surviving daughters)	surviving daughters of females in col. 1 born 1940/55
0–14	14 (A)	16 (P)	4 (a)
15–29	15 (B)	13 (Q)	10 (b)
30–44	11 (C)	14 (R)	2 (c)

group during the period 1940 to 1955. The second column shows the numbers of the original women and daughters surviving in 1955 in the three age groups. Could we have predicted the numbers for 1972? The entries in each column are lettered for easy reference. We should notice first that a, b, and c, all born since 1940, must total to P, that is:

$$a + b + c = P \qquad (1)$$

The birth rate coefficients for classes A, B, and C are obviously a/A, b/B, and c/C. We assume these to be constant in the period under review.

Our problem now is this. Given these figures, can we construct a matrix, M, for predicting the female population structure in, say, 1972— or even 1985? Now such a matrix must 'predict' P, Q, and R from A, B, and C. Mathematicians would say that M must map the column vector:

$$\begin{pmatrix} A \\ B \\ C \end{pmatrix} \quad \text{into} \quad \begin{pmatrix} P \\ Q \\ R \end{pmatrix}$$

Suppose that:

$$M = \begin{pmatrix} p & q & r \\ s & t & u \\ v & w & x \end{pmatrix}$$

Then:

$$\begin{pmatrix} p & q & r \\ s & t & u \\ v & w & x \end{pmatrix} \begin{pmatrix} A \\ B \\ C \end{pmatrix} = \begin{pmatrix} P \\ Q \\ R \end{pmatrix}$$

So that:

$$pA + qB + rC = P \qquad (2)$$

If we take $p = a/A$, $q = b/B$ and $r = c/C$, equation (2) reduces to equation (1). Next, Q can only consist of A multiplied by A's survival rate Q/A. Hence $s = Q/A$, $t = 0$, and $u = 0$. Finally, R can only consist of B multiplied by B's survival rate R/B. Hence $w = R/B$, $v = 0$, and $x = 0$. It follows, therefore, that our prediction matrix must be:

$$\begin{pmatrix} \dfrac{a}{A} & \dfrac{b}{B} & \dfrac{c}{C} \\[2ex] \dfrac{Q}{A} & 0 & 0 \\[2ex] 0 & \dfrac{R}{B} & 0 \end{pmatrix}$$

With our figures this is:

$$\begin{pmatrix} \frac{2}{7} & \frac{2}{3} & \frac{2}{11} \\ \frac{13}{14} & 0 & 0 \\ 0 & \frac{14}{15} & 0 \end{pmatrix}$$

A check shows that this does indeed transform:

$$\begin{pmatrix} 14 \\ 15 \\ 11 \end{pmatrix} \text{ into } \begin{pmatrix} 16 \\ 13 \\ 14 \end{pmatrix}$$

We can now use it to predict the female population figure for 1972:

$$\begin{pmatrix} \frac{2}{7} & \frac{2}{3} & \frac{2}{11} \\ \frac{13}{14} & 0 & 0 \\ 0 & \frac{14}{15} & 0 \end{pmatrix} \begin{pmatrix} 16 \\ 13 \\ 14 \end{pmatrix} = \begin{pmatrix} 15.8 \\ 14.9 \\ 12.1 \end{pmatrix}$$

The reader may want to predict the 1985 figures. The result, of course, will be far less reliable than our prediction for 1970 since our assumption about fixed birth coefficients and survival rates can only apply over a limited period. Indeed, the way in which these factors change and alter the structure of society is a study in itself. A paper delivered to the 1969 Institute of Mathematics and its Applications Computer Symposium at Girton College, Cambridge, by R. S. Schofield of the Cambridge Group for the History of Population and Social Structure, describes current research into family reconstitution by computer linking of vital records. In the process, every conceivable item of information about a given family is extracted from parish registers—dates of baptism, marriage and burial. Where sufficient information is available it is sorted and matched with that of related families to produce demographic indices of fertility, mortality, and nuptiality. The work is important to those researching trends in sociological attitudes and to geneticists alike.

Solving the problems

Assessing accurately the future sizes and needs of populations is an essential part of the solution. Until a problem is clearly defined, attempts to solve it are valueless. But the problem itself is a complex of many smaller problems, and solving one may aggravate others or even create new difficulties. In the past, when man has overeaten, he has become a prey to disease; nature has imposed its own cruel, pragmatic solutions; but when disease is controlled, over-population and starvation ensue. Indeed, the averting of the First and Second World Wars would undoubtedly have saved the lives of many, but it would have increased the poverty and misery of others. In the era of nuclear warfare such man-made checks as world war are unimaginable. The only alternatives are to limit population increase by voluntary artificial means, and to improve the technology of food production.

Progress is now under way in food production technology. In Asia, strains of rice already exist which yield twice in a season. Some of the Indian institutes of technology are engaged in intensive research to develop new seed, to determine optimum conditions for existing crops, and to increase the efficiency of agricultural machinery. Meanwhile, substantial help to developing countries flows from the West, much of it through United Nations Organizations such as UNESCO and UNICEF. In addition, money is provided for equipment and training support for horticulture, poultry, and pisciculture development programmes.

Despite all this we sense that we are only at the beginning of developing our natural resources. Let us consider the sources of food and minerals in the oceans. In actual practice only 10% of the world's animal protein and 1% of the world's total food supplies come from the sea, and yet the sea covers 70% of the Earth's surface—139 million square miles (360 million sq km)—and its average depth is 12 500 feet (3 800 m). Indeed, water itself is essential to life and industry. In the West, the average consumption per head is 50 gallons (230 litres) per day, though in the developing countries more like 3 gallons or 14 litres per head are used. It is estimated that present needs will be doubled by 1985. Now 97% of all the water in the world is in the sea, and a further 2% is captive in the ice caps and glaciers; only 1% of the whole is actually in circulation.

The year 2000

We are not without star-gazers and soothsayers to foretell the future, and, indeed, many societies and groups are actually engaged in really business-like forecasting of its problems. In Britain the Social Science Research Council has formed a Committee on the next Thirty Years. In the United States a Resources for the Future organization has been set up with assistance from the Ford Foundation, and the American Academy of Arts and Science has established a Commission on the

Year 2000. This has groups working on such studies as 'Government and Structure', 'Values and Rights', 'The Life Cycle of the Individual', 'Social Consequences of the Computer', 'Problems of Biomedical Engineering', and 'The future Organization of Science and Technology'. The Hudson Institute and the Rand Corporation have both sponsored similar studies and, of course, there are investigations by individuals, such as Hermann Landsberg, Leonard Fischmann and Joseph Fisher's 'Resources in America's Future' (1963). In France there are Gaston Berger's *Prospectives* group and Bertrand de Jouvenal's *Futuribles* project.

In their now classic horoscopes in *The Year 2000*, Kahn and Wiener discuss the long-term trends of Western society. They see, of course, population growth, increasing urbanization and the development of megalopolises. The accumulation of scientific and technical knowledge will not only speed up the tempo and institutionalization of change and the world-wide spread of industrialization and modernization, but will lead to more affluence and leisure and a decreasing importance of primary and even secondary occupations. Literacy and education will inevitably increase, but whether sufficiently to enable man to control his greatly increased powers of self-destruction is an open question. Kahn and Wiener see men generating increasingly sensate cultures and encouraging the development of meritocratic, bureaucratic and nationalistic *élites* within the structures of government. Perhaps there is comfort to be had from Machiavelli's dictum that half men's actions are ruled by chance and the other half by men themselves.

If man can, by his own wit or good fortune, survive beyond this, he must emerge into a 'post-industrial' era—in which business is no longer the principal motivating source of innovation; in which most economic activities have become 'service oriented' instead of 'production oriented'; in which the average per capita income will be fiftyfold that of the industrial era, and in which the market will play a diminished role compared with the public sector accounts. Cybernation established, with computers harnessed to national record-keeping and routine decision-making, to the ordering and classification of scientific research, to banking, medical diagnosis, traffic control, crime control and prevention, weather forecasting, and so on, we may revert to a learning society—really valuing and extending our educational institutions, at last losing our rat race complexes and reducing our nationalistic tendencies.

Even so, world government is still not yet achieved, political power blocs will remain and by the 'post-industrial' era these may include Japan, China, a European complex, Brazil or Mexico, and perhaps India. Such long-distance crystal-gazing presupposes a considerable degree of success with population control, arms control and international security generally. In case these glimpses of Canaan beguile us into forgetting that, far from being solved, the population problem looms more menacingly with every day that passes, we conclude with a passage from Lord Ritchie-Calder's apocalyptic lecture 'Hell on Earth' given to the Conservation Society in 1968 in London.

'On present trends a child born today will see within its lifetime the emergence of cities with over 1000 million inhabitants. Of the 7000 million (on present trends) by the end of the century and the 15 000 million by the year 2045 (when children born this year will be 76), a growing proportion will live in cities. At present 38% of the world's population is urban, but the move from the countryside could rèach 50% in 16 years and 100% within 55 years. During the lifetime of children born today the whole of mankind could be city bound. The biggest city would have 1300 million inhabitants (186 times the present population of Greater London).

'There has always been overpopulation and overexploitation. Many past civilizations have buried themselves in the graveyards of their own mistakes—but others rose to replace them because civilizations were linked to a particular river basin or region. But today we have a global civilization. It is a community so interdependent that every mistake we make is exaggerated on a world scale.

'Population figures tend to produce "statistical anaesthesia", but if you have ever seen a dead infant taken from a mother's empty breasts you won't forget—ever—what hunger and famine mean.'

Further Reading

Chapter 1: The mind of the mathematician

BELL, E. T. *Men of Mathematics* (Harmondsworth: Penguin Books, 1965).

BIXBY, WILLIAM, (with the editors of *Horizon* Magazine) *The Universe of Galileo and Newton* (London: Cassell, 1966).

BUTTERFIELD, HERBERT *The Origins of Modern Science* (New York: Collier Books, 1962).

CARROLL, LEWIS *Alice's Adventures in Wonderland* and *Through the Looking Glass* (London: Macmillan, 1912).

COURANT, RICHARD and ROBBINS, HERBERT *What is Mathematics?* (London and New York: Oxford University Press, 1967).

DANTZIG, TOBIAS *Number, the Language of Science* (London: Allen and Unwin, 1930).

EINSTFIN, ALBERT *Relativity*, translated by Robert W. Lawson (London: Methuen, 1921).

FARBER, S. M., and WILSON, R. H. L. (editors) *Control of the Mind* (New York and London: McGraw-Hill, 1961).

GEYMONAT, LUDOVICO *Galileo Galilei* (New York: McGraw-Hill, 1965).

HADAMARD, JACQUES *The Psychology of Invention in the Mathematical Field* (Princeton University Press, 1949).

HOLT, MICHAEL Letter to *Science* **146**, no. 3647, Washington DC: November 1964, p. 998.

KASNER, EDWARD, and NEWMAN, J. R. *Mathematics and the Imagination* (London: Bell, 1949).

KILMISTER, C. W., *Language, Logic and Mathematics* (London: English Universities Press, 1967).

KOESTLER, ARTHUR 'The Greatest Scandal in Christendom', *Observer*, London, 2 February, 1964.

NEWMAN, JAMES R. (editor) *The World of Mathematics* (London: Allen and Unwin, 1961).

NEWMAN *Science and Sensibility* (New York: Simon and Schuster, 1961).

POINCARE, HENRI *The Foundations of Science* (Pennsylvania : The Scie ce Press, 1929).

POLYA, GEORGE *Mathematical Discovery* (New York: John Wiley, 1965), vol. II.

RUSSELL, BERTRAND, *The Principles of Mathematics* (London: Allen and Unwin, 1903).

RUSSELL *History of Western Philosophy* (London: Allen and Unwin, 1961).

STEIN, SHERMAN K. *Mathematics, the Man-made Universe* (San Francisco and London: W. H. Freeman, 1963).

WEINBERG, ALVIN M. *Reflections on Big Science* (Oxford: Pergamon, 1967).

WERTHEIMER, MAX *Productive Thinking* (Harper and Brothers, 1945).

Chapter 2: Thinking and reasoning

ALLEN, LAYMAN E. 'Games and Programmed Instruction', paper for Conference on the Implications of Developments in the Communication Sciences for Legal Education in the Next Decade, September 1963.

BERGSON, HENRI, *Creative Evolution* (Trans. A. Mitchell (Basingstoke: Macmillan, 1970)).

BEVERIDGE, W. T. *The Art of Scientific Investigation* (London: Heinemann, 1968).

BRUNER, JEROME S. 'The Nature of Learning in Childhood', paper presented to Rehovot Conference of Science and Education in Developing States, August 1969.

BRUNER, GOODNOW J. J., and AUSTIN, G. A. *A Study of Thinking* (New York: John Wiley, 1956).

DAVY, JOHN 'The Chomsky Revolution', *Observer*, 10 August 1969.

DE BONO, EDWARD *The Use of Lateral Thinking* (London: Cape, 1967).

DIENES, ZOLTAN P. 'On Abstraction and Generalization', *Harvard Educational Review*, Summer 1961.

DIENES *An Experimental Study of Mathematics Learning* (London: Hutchinson, 1963).

DIENES 'On the Learning of Mathematics', *The American Teacher*, March 1963.

DIENES 'Insight into Arithmetical Processes', *The School Review*, The University of Chicago Press, vol. 72, no. 2, 1964.

DIENES 'Some Basic Processes Involved in Mathematics Learning', *Research in Mathematics Education* (National Council of Teachers of Mathematics, Inc., 1967).

DIENES and JEEVES *The Effects of Structural Relations on Transfer* (London: Hutchinson, 1970).

ENGEL, BERNARD and KAMYA, JOE Reported in *Time*, 18 July 1969.

FISCHER, ROLAND 'The Perception–Hallucination Syndrome (A Re-Examination)', *Diseases of the Nervous System*, Royal Society of Medicine, March 1969, vol. 30, pp. 161–71.

FLESCH, RUDOLF *Why Johnnie Can't Read* (New York: Harper and Row, 1950).

VON FOERSTER, HEINZ 'What is Memory that it may have Hindsight and Foresight as well', *Proceedings of Third International Conference of the Future of Brain Science*, New York, May 1968.

GATTEGNO, C., *et al. Le Materiel pour l'enseignement des mathematiques* (Paris: Neuchatel, 1958).

GOMBRICH, E. H. *Art and Illusion* (London: Phaidon Press, 1962).

GREGORY, R. L. *Eye and Brain* (London: Weidenfeld and Nicolson, 1966).

GUILFORD, J. P. *The Nature of Intelligence* (New York: McGraw-Hill, 1967).

HABER, RALPH NORMAN 'Eidetic Images' *Scientific American*, April 1969.

HOCHBERG, JULIAN E. *Perception* (Englewood Cliffs, New Jersey: Prentice-Hall, 1964).

HOLT, MICHAEL *Mathematics in Art* (London: Studio Vista, 1971).

HOLT and DIENES *Let's Play Maths* (Harmondsworth: Penguin Books, 1972).

HUG, COLETTE *L'Enfant et la Mathematique* (Paris: Bordas-Mouton, 1968).

JEEVES, M. A. and DIENES *Thinking in Structures* (London: Hutchinson, 1968).

KLINE, MORRIS (ed.) *Mathematics in the Modern World* (Scientific American/ W. H. Freeman, 1968).

KOESTLER, ARTHUR *The Act of Creation* (London: Hutchinson, 1964).

KOESTLER and SMYTHIES, J. R. (editors) *Beyond Reductionism* (London: Hutchinson, 1969).

PAVLOV, P. *Conditioned Reflexes* (London: Oxford University Press, 1927).

PIAGET, JEAN *The Child's Conception of Number* (London: Routledge and Kegan Paul, 1952).

PIAGET and INHELDER, BARBEL, 'The Gaps in Empiricism', in Koestler and Smythies, *Beyond Reductionism*.

PIAGET *The Mechanisms of Perception* (New York: Basic Books Inc., 1969).

POPPER, KARL *The Logic of Scientific Discovery* (London: Hutchinson, 1959).

ROKEACH, MILTON *The Open and Closed Mind* (New York: Basic Books, 1960).

ROSENBLOOM, PAUL C. 'A Mathematical Game of Golf', Minnesota University MINNEMAST Mathematics Materials, 1964.

ST AUGUSTINE *St Augustine's Confessions* translated by William Watts, 1631 Loeb edition (London: William Heinemann, 1912).

SHEO DAN SINGH 'Urban Monkeys', *Scientific American*, July 1969, vol. 221, no. 1.

SUPPES, PATRICK 'Modern Learning Theory and the Elementary-School Curriculum', *American Educational Research Journal*, March 1964, vol. 1, no. 2.

SUPPES and IHRKE, CONSTANCE 'Accelerated Program in Elementary-School Mathematics—The Third Year', Technical Report no. 108, January 1947, Institute of Mathematical Studies, Stanford University.

SUPPES, and BINFORD, FREDERICK 'Experimental Teaching of Mathematical Logic in the Elementary School', *The Arithmetic Teacher*, March 1965.

SUPPES and HANSEN, DUNCAN 'Accelerated Program in Elementary-School Mathematics—The First Year', *Psychology in the Schools*, 1965, no. 2, pp. 195–203.

SUPPES, HYMAN, LESTER, and JENMAN, MAX, 'Linear Structural Models for Response and Latency Performance in Arithmetic', Technical Report no. 100, 29 July 1966, Institute of Mathematical Studies, Stanford University.

SUPPES 'Accelerated Program in Elementary-School Mathematics—'The Second Year', *Psychology in the Schools*, October 1966, vol. III, no. 4, pp. 294–307.

SUPPES, and GREEN, GUY 'Some Counting Models for First Grade Performance Data on Simple Addition Facts', *Research in Mathematics Education* (National Council of Teachers of Mathematics, Inc., 1967).

VIAUD, GASTON *Intelligence, its Evolution and Form* (London: Arrow Books, 1960).

WHITEHEAD, A. N. *Science and the Modern World* (Cambridge University Press, 1927).

WITKIN, HERMAN A. and OLTMAN, PHILIP K. 'Cognitive Style', *International Journal of Neurology*, 1967, vol. 6, no. 2, pp. 119–34.

YOUNG, J. Z. 'The Organization of a Memory System', *Proceedings of the Royal Society*, **B 163,** The Croonian Lecture (1965), pp. 285–320.

Chapter 3 The unreasonableness of logic

BRUNER et al., *A Study of Thinking* (New York: John Wiley, 1956).

CARAMAN, PHILIP *John Gerard* (London: Longmans, Green, 1965).

LUKASIEWICZ, JAN, *Elements of Mathematical Logic* (Oxford: Pergamon Press, 1963).

Chapter 4: Models in science

EUCLID *The Elements*, Book 9, Proposition 20.

FEYNMAN, RICHARD *The Character of Physical Law*, book of the televised Messenger lectures (BBC, 1965).

FEYNMAN *Lectures on Physics*, vols 1 and 2 (Reading, Massachusetts: Addison-Wesley, 1963).

VON FOERSTER, HEINZ 'From Stimulus to Symbol: the Economy of Biological Computation', *Sign Image and Symbol*, Gyorgy Kepes (ed.) (London Studio Vista, 1966).

HARDY, G. H. *A Mathematician's Apology* (Cambridge University Press, 1970).

HORADAM, A. F. *Outline Course of Pure Mathematics* (London: Pergamon Press, 1968).

RUSTOGI, J. S. *The Mathematics Seminar of India*, Vol. 5, no. 2, 1968.

WATKINS, J. W. N. 'Confession is Good for Ideas', *Experiment*, Broadcasts for BBC Third Programme (BBC, 1964).

WATSON, JAMES D. *The Double Helix* (London: Weidenfeld and Nicolson, 1968).

WILDE, OSCAR *The Importance of Being Earnest*, first publ. 1895 (London: Heinemann Educational Books, 1949).

Chapter 5: The mathematics of mind

BRONOWSKI, J., and MAZLICH, BRUCE *The Western Intellectual Tradition* (Harmondsworth: Penguin Books, 1963).

DE BONO, EDWARD *The Mechanism of Mind* (London: Cape, 1969).

FEYNMAN, RICHARD *Lectures on Physics*, vol. 1 (Reading, Massachusetts: Addison-Wesley, 1963).

KOESTLER, ARTHUR, and SMYTHIES, J. R. (eds.) *Beyond Reductionism* (London: Hutchinson, 1969).

WADDINGTON, C. H. 'The Theory of Evolution Today', in *Beyond Reductionism* (London: Hutchinson, 1969).

ZEEMAN, E. C. 'Topology of the Brain', *Mathematics and Computer Science in Biology and Medicine* (London: HMSO, 1965).

ZEEMAN 'The Brain is no Computer', *Planar News*, January 1968, and *Pulse*, 28 September 1968.

ZEEMAN 'A Mathematical Explanation of Creativity', from *The Brain*, a BBC Radio 3 Series (1968).

ZEEMAN and BUNEMAN, O. P. 'Tolerance Spaces and the Brain', *Towards a Theoretical Biology*, ed. C. H. Waddington (London: English Universities Press, 1967).

Chapter 6: Economy

BATTERSBY, A. *Mathematics in Management* (Harmondsworth: Penguin Books, 1966).

BAUMOL, W. J. *Economic Theory and Operations Analysis*, 2nd edition, chap. 20 (Englewood Cliffs, New Jersey: Prentice-Hall, 1965).

BELL, E. T. *Men of Mathematics* 2 (Harmondsworth: Penguin Books, 1965).

BOEHM, GEORGE A. W. *The New World of Mathematics* (London: Faber and Faber, 1959).

BUSACKER, R. G. and SAARTY, T. L. *Finite Graphs and Networks* (New York: McGraw-Hill, 1965).

CAMERON, B. *Input/Output Analysis and Resource Allocation* (London: Oxford University Press, 1968).

CAYLEY, A. 'A Memoir on the Theory of Matrices', *Phil. Trans. Royal Society*, 1858.

CHIANG, A. C. *Fundamental Methods of Mathematical Economics* (New York: McGraw-Hill, 1967).

CONWAY, FREDA 'Patterns of Household Expenditure', *Mathematical Gazette*, October 1968.

DAVIS, MORTON D. *Game Theory* (New York: Basic Books Inc., 1970).

DORFMAN, R. *et al. Linear Programming and Economic Analysis* (New York: McGraw-Hill, 1958).

FLETCHER, T. J. 'Combining Matrices', *Mathematical Gazette*, February 1968.

GLENN, J. 'Mathematics and Economics Theory', *Mathematics Teaching*, Autumn 1970.

KAHN, K. and WIENER, A. *The Year 2000.*

KEMENY, J. G., SCHLEIFER, SNELL and THOMPSON *Finite Mathematics with Business Application* (Englewood Cliffs, New Jersey: Prentice-Hall, 1963).

KEYNES, J. M. *The General Theory of Employment, Interest and Money* (Harmondsworth: Penguin Books, 1967).

LEKACHMAN, ROBERT *The Age of Keynes* (Harmondsworth: Penguin Books, 1969).

LEONTIEF, W. W. *Input–Output Economics* (New York: Oxford University Press, 1966).

MURPHY, BRIAN *The Computer in Society* (London: Blond, 1967).

VON NEUMANN, JOHN and MORGENSTERN, OSKAR *The Theory of Games and Economic Behaviour* (2nd edition) (Princeton: John Wiley, 1953).

PEN, J. *Modern Economics* (Harmondsworth: Penguin Books, 1970).

SAMUELSON, P. A. *Economics* (New York: McGraw-Hill, 1970).

VAJDA, S. *The Theory of Games and Linear Programming* (London: Methuen, 1956).

WILSON, ANDREW *The Bomb and the Computer* (London: Barrie and Jenkins, 1968).

Chapter 7 Mathematics and maps

BICKMORE, D. P. 'Maps for the Computer Age', *Geographical Magazine*, December 1968.

BIRCH, T. W. *Maps* (Oxford University Press, 1964).

BOOKER, P., FREWER, G., and PARDOE, D. *Project Apollo—The Way to the Moon* (London: Chatto and Windus, 1969).

BOWYER, R. E. and GERMAN, G. A. *A Guide to Map Projections* (London: John Murray, 1959).

'Cartography for the 70's', (a whole issue devoted to modern techniques in map-making), *Geographical Magazine*, October 1969.

CAYLEY, G. 'On the Colouring of Maps', *Proceedings of the Royal Geographical Society*, **1**, 259–61.

CHUBB, T. *The Printed Maps in the Atlases of Great Britain and Ireland 1579/1880* (Folkestone: Dawsons of Pall Mall, 1966).

COTTER, C. H. *Astronomical and Mathematical Foundations of Geography* (London: Hollis and Carter, 1966).

FIELDER, G. *Structure of the Moon's Surface* (Oxford: Pergamon Press, 1961).

HOOD, P. *How Time is Measured* (Oxford University Press, 1969).

LINTON, D. L. 'A View of the Craters of the Moon', *Geographical Magazine*, April 1969.

LISTER, R. *How to Identify Old Maps and Globes* (London: Bell, 1965).

'Man on the Moon', *Science Journal*, May 1969.

MOORE, PATRICK *Sky at Night*, vol. 1 and 2 (Eyre and Spottiswood, 1964 and 1968), vol. 3 (BBC, 1970).

MOORE *Space* (Woking: Lutterworth Press, 1970).

MOORE *Moon Flight Atlas* (London: George Philip, 1970).

RADFORD, P. J. *Antique Maps* (Portsmouth: Radford, 1965).

RAISS, ERWIN *General Cartography* (New York: McGraw-Hill, 1948).

ROBINSON, A. H. and SALE, R. D. *Elements of Cartography* (New York: John Wiley, 1969).

ROBLIN, H. S. *Map Projections* (London: Edward Arnold, 1969).

TOOLEY, R. V. *Maps and Map Making* (London: Batsford, 1970).

WARTNABY, J. *Surveying* (London: HMSO).

Chapter 8: Shapes and sizes in geography

AMBROSE, P. *Analytical Human Geography* (Harlow: Longmans, 1969).

BERRY, B. J. L. and MARBLE, D. F. *Spatial Analysis* (Englewood Cliffs, New Jersey: Prentice-Hall, 1968).

BUSACKER, R. G. and SAATY, T. L. *Finite Graphs and Networks* (New York: McGraw-Hill, 1965).

CHISHOLM, M. *Rural Settlement and Land Use* (London: Hutchinson, 1966).

CHORLEY, R. J. and HAGGETT, P. *Socio-economic Models in Geography* (London: Methuen, 1968).

CHORLEY, and HAGGETT *Models in Geography* (London: Methuen, 1967).

CHORLEY and HAGGETT 'Trend Surface Mapping in Geographical Research', *Institute British Geographers Publications 37*, pp. 47–67.

COLE, J. P. *Bulletins of the Department of Geography* (Nottingham University).

No. 2 'Set Theory and the Geography'

No. 8 'An Introduction to Factor Analysis with the US as an example'

No. 10 'Topology and Geography'

No. 11 'Measurement of Shape in Geography'

No. 12 'An Introduction to Trend Surface Analysis'

No. 16 'The Population of S. America from 1870–1960 by Major Civil Divisions'

No. 17 'Places in the News, a Study of Geographical Information'

Education Pamphlet No. 59 *New Thinking in Geography* (London: HMSO, 1972).

ESTALL, R. C. and BUCHANAN, R. O. *Industrial Activity and Economic Geography* (London: Hutchinson, 1966).

EVERSON, J. A. and FITZGERALD, *Settlement Patterns* (London: Longmans, 1969).

HAGGETT *Locational Analysis in Human Geography* (London: Arnold, 1965).

KANSKY, K. J. 'Structure of Transport Networks: relationships between network geometry and regional characteristics', *University of Chicago Dept. of Geog., Research Papers* 84.

KING, C. A. M. *Mathematical Education in Science and Technology*, vol. 1, no. 2.

SMITH, R. H. T. *et al. Reading in Economic Geography* (Chicago: Rand McNally).

TOULMIN, S. E. *The Philosophy of Sciences*

THEAKSTONE, W. H. and HARRISON, C. *The Analysis of Geographical Data* (London: Heinemann, 1970).

Chapter 9: Population

ASHBY, W. R. *An Introduction to Cybernetics* (London: Methuen).

BEAUJEU-GARNIER, J. *Geography of Population* (Harlow: Longmans, 1966).

BRIERLEY, J. K. *Biology and the Social Crisis* (London: Heinemann Educational Books, 1967).

BRIERLEY *A Natural History of Man* (London: Heinemann Educational Books 1970).

COALE, A. J., and HOOVER, E. M. *Population Growth and Economic Development in Low-Income Countries* (Princeton University Press, 1958).

COX, P. R. *Demography* (Cambridge University Press, 1959).

DARWIN, CHARLES *On the Origin of Species*, Everyman Library (London: Dent, 1956).

Demographic Year Book (United Nations, 1963).

The Determination and Consequences of Population Trends, Report no. 17, Population Branch of the Department of Social Affairs, United Nations.

FAWCETT, C. B. 'Population Maps', *Geog. Journal*, February 1935, pp. 142–59.

FISHER, RONALD *Genetical Theory of Natural Selection* (Dover Publications, 1958).

FOX, R. *Kinship and Marriage* (Harmondsworth: Penguin Books, 1970).

Great World Atlas (Readers Digest) (London: Hodder, 1968).

HALDANE, J. B. S. *The Causes of Evolution* (Cornell University Press, 1968).

HALDANE 'Mathematics of Natural Selection' in *The World of Mathematics*, vol. 2, Ed. by J. R. Newman (London: Allen and Unwin, 1960).

KAHN, HERMAN and WIENER, ALBERT *The Year 2000* (New York: Collier-Macmillan, 1967).

LAZARSFIELD, P. F. *et al. Readings in Mathematical Social Science* (Chicago: MIT Press, 1969).

LI, C. C. *Population Genetics* (University of Chicago Press, 1955).

MALTHUS, ROBERT (ed. Few, A.) *An Essay on the Principle of Population as it Affects the Future Improvements of Society 1798* (Harmondworth: Penguin, 1971).

MARRIOTT, F. H. C. *Basic Mathematics for the Biological and Social Sciences* (Oxford: Pergamon Press, 1970).

MORRIS, DESMOND *The Naked Ape* (London: Cape, 1967).

MORRIS *The Human Zoo* (London: Cape, 1971).

SEARLE, S. R. *Matrix Algebra for the Biological Sciences* (New York: John Wiley, 1966).

SLOBODKIN, L. B. *Growth and Regulation of Animal Populations* (New York: Holt, Rinehart and Winston, 1961).

SMITH, J. MAYNARD *Mathematical Ideas in Biology* (Cambridge University Press, 1968).

SOLOMON, M. E. *Population Dynamics* (London: Edward Arnold, 1969).

ZELINSKY, W. *A Prologue to Population Geography* (London: Prentice Hall, 1970).

Index